Arduino Essentials

Enter the world of Arduino and its peripherals and start creating interesting projects

Francis Perea

PUBLISHING
BIRMINGHAM - MUMBAI

Arduino Essentials

Copyright © 2015 Packt Publishing

All rights reserved. No part of this book may be reproduced, stored in a retrieval system, or transmitted in any form or by any means, without the prior written permission of the publisher, except in the case of brief quotations embedded in critical articles or reviews.

Every effort has been made in the preparation of this book to ensure the accuracy of the information presented. However, the information contained in this book is sold without warranty, either express or implied. Neither the author, nor Packt Publishing, and its dealers and distributors will be held liable for any damages caused or alleged to be caused directly or indirectly by this book.

Packt Publishing has endeavored to provide trademark information about all of the companies and products mentioned in this book by the appropriate use of capitals. However, Packt Publishing cannot guarantee the accuracy of this information.

First published: February 2015

Production reference: 1200215

Published by Packt Publishing Ltd.
Livery Place
35 Livery Street
Birmingham B3 2PB, UK.

ISBN 978-1-78439-856-9

www.packtpub.com

Credits

Author
Francis Perea

Reviewers
Ladislas de Toldi

Ken Leung

Commissioning Editor
Sarah Crofton

Acquisition Editor
Meeta Rajani

Content Development Editor
Ajinkya Paranjpe

Technical Editor
Sebastian Rodrigues

Copy Editors
Nithya P

Stuti Srivastava

Project Coordinator
Harshal Ved

Proofreaders
Maria Gould

Jonathan Todd

Bernadette Watkins

Indexer
Monica Ajmera Mehta

Production Coordinator
Shantanu N. Zagade

Cover Work
Shantanu N. Zagade

About the Author

Francis Perea is a vocational education professor at Consejería de Educación de la Junta de Andalucía in Spain with more than 14 years of experience.

He has specialized in system administration, web development, and content management systems. In his spare time, he works as a freelancer and collaborates, among others, with ñ multimedia, a little design studio in Córdoba, working as a system administrator and the main web developer.

He has also collaborated as a technical reviewer on *SketchUp 2014 for Architectural Visualization Second Edition*, *Arduino Home Automation Projects*, and *Internet of Things with the Arduino Yún*, all by Packt Publishing.

When not sitting in front of a computer or tinkering in his workshop, he can be found mountain biking, kite surfing, or working as a beekeeper, taking care of his hives in Axarquía County, where he lives.

To Salomé: I owe it all to you.

About the Reviewer

Ladislas de Toldi is a biotech engineer who has always been passionate about computers, robotics, and artificial intelligence.

Ladislas is the CEO at Leka, a young start-up whose goal is to use robotics to help exceptional children live a normal life.

He has been working on Arduino for several years and has contributed to several open source projects, such as Sudar's Arduino-Makefile (`https://github.com/sudar/Arduino-Makefile`).

www.PacktPub.com

Support files, eBooks, discount offers, and more

For support files and downloads related to your book, please visit www.PacktPub.com.

Did you know that Packt offers eBook versions of every book published, with PDF and ePub files available? You can upgrade to the eBook version at www.PacktPub.com and as a print book customer, you are entitled to a discount on the eBook copy. Get in touch with us at service@packtpub.com for more details.

At www.PacktPub.com, you can also read a collection of free technical articles, sign up for a range of free newsletters and receive exclusive discounts and offers on Packt books and eBooks.

https://www2.packtpub.com/books/subscription/packtlib

Do you need instant solutions to your IT questions? PacktLib is Packt's online digital book library. Here, you can search, access, and read Packt's entire library of books.

Why subscribe?

- Fully searchable across every book published by Packt
- Copy and paste, print, and bookmark content
- On demand and accessible via a web browser

Free access for Packt account holders

If you have an account with Packt at www.PacktPub.com, you can use this to access PacktLib today and view 9 entirely free books. Simply use your login credentials for immediate access.

Table of Contents

Preface	**1**
Chapter 1: Meeting the Arduino Family	**7**
A game changer	**7**
Common features	**8**
Arduino Uno	9
Arduino Mega 2560	11
Arduino Ethernet	12
LilyPad Arduino	14
Arduino Yún	16
Arduino Mini, Micro, and Nano	18
Other Arduino family members	19
Esplora	19
Arduino Robot	20
Arduino Due	22
Unofficial boards	22
Shields	23
Just one to rule them all	**24**
Users teaching users	**24**
Summary	**25**
Chapter 2: The Arduino Development Environment	**27**
A multiplatform tool	**27**
Downloading the package	**28**
Windows	28
Mac OS X	28
Linux	29
Source code	29

Table of Contents

Installing the software	**29**
Windows	29
Mac OS X	29
Linux	30
In case of trouble	30
Installing the drivers	**30**
Windows	31
Mac OS X	32
Linux	32
Running the Arduino development environment for the first time	**32**
The toolbar	34
The code editor	34
The message area	36
Preflight checks	**36**
Uploading our first sketch	**38**
Main menus and commands	**40**
The Arduino language	**42**
Summary	**42**
Chapter 3: Interacting with the Environment the Digital Way	**43**
Digital versus analog signals	**43**
Our first circuit	**44**
Using a breadboard	45
The LED circuit	46
Circuit schematic	47
Breadboard connections diagram	48
Asymmetric blinking code	49
C language syntax considerations	52
Troubleshooting faults in the circuit	52
Dealing with multiple outputs	53
Current limit per pin	55
Summary	**59**
Chapter 4: Controlling Outputs Softly with Analog Outputs	**61**
Dealing with analog signals	**61**
The analog output circuit	**62**
Connections diagram	**63**
Analog control through code	**64**
The analogWrite() function	64
The for loop	64
Complete the fading LED code	66

Motor control with a transistor	68
Motor driver	69
Power source considerations	70
The complete circuit	71
Connections diagram	72
Motor varying speed code	73
The assembled circuit	74
Bigger power motors	74
Summary	**76**
Chapter 5: Sensing the Real World through Digital Inputs	**77**
Sensing by using inputs	77
Connecting a button as a digital input	78
The momentary push button	80
Complete circuit schematic	82
Breadboard connections diagram	83
Writing code to react to a press	84
Configuring and reading a digital input	85
Taking decisions with conditional bifurcations	86
Timing and debouncing	87
Other types of digital sensors	87
Using an optocoupler as a coin detector	89
The schematic of the coin detector	89
The breadboard connections diagram	90
The complete example code	92
A real working project	93
Summary	94
Chapter 6: Analog Inputs to Feel Between All and Nothing	**95**
Sensing analog values	95
The Arduino map function	96
An ambient light meter	97
Connecting a variable resistor to Arduino	98
Voltage divider	99
An ambient light meter circuit	99
Breadboard connections	100
Programming to sense the light	101
An ambient light meter code	101
The switch / case control structure	102
Calibrating the sensor	105

Table of Contents

DC motor speed control revisited	**105**
The potentiometer	106
The motor speed control schematic	106
The breadboard connections diagram	107
A simple code to control the motor speed	108
Summary	**110**
Chapter 7: Managing the Time Domain	**111**
Time control functions	**111**
Stopping versus accounting	112
Making some noise	**112**
Arduino library sound functions	113
Sound hardware connection	114
Direct connection	115
Connection through a transistor	116
A simple timer	118
Dividing your sketch into different files	118
Coding a timer by using delays	119
Coding without delays and blinking an LED while waiting	121
A bigger project – a metronome	**124**
The metronome circuit	125
The metronome code	126
Summary	**129**
Chapter 8: Communicating with Others	**131**
Serial communications concepts	**131**
The baud rate	133
Other types of serial communication	**133**
Calibrating sensors serially	**134**
Sending data to Arduino	**138**
A computer connected dial thermometer	**142**
The thermometer circuit	143
The code for the thermometer	145
Summary	**148**
Chapter 9: Dealing with Interrupts	**149**
The concept of an interruption	**149**
The ISR	150
The tachograph project	**151**
Mechanical considerations	151
A simple interrupt tester	152
Our first interrupt and its ISR	154

A dial tachograph	**156**
Breadboard connections diagram	157
The complete tachograph code	158
Modular development	164
Summary	**165**
Chapter 10: Arduino in a Real Case – Greenhouse Control	**167**
A greenhouse controller	**167**
The controller requirements	168
Modular design	168
Temperature control	169
Humidity control	169
Lighting control	169
Manual alarm	170
Input and output devices	170
The relay as a mediator	171
The greenhouse controller circuit	173
Breadboard connections diagram	175
The greenhouse controller code	**176**
Libraries and constant definitions	176
Global variables	177
The interrupt ISR	178
The alarm routine	178
Initialization and board configuration	180
The main execution loop	180
Temperature subsystem	180
Humidity subsystem	181
Lighting subsystem	182
Alarm subsystem	182
Serial feedback and calibration	183
The complete project code	183
Final considerations	183
Summary	**184**
Index	**185**

Preface

The Arduino platform has become a de facto standard when talking about microcontrollers. With a wide range of different board models, it can cover a wide spectrum of projects, and its ease of use has made it the preferred platform for those starting out in the microcontroller world.

If you are a hobbyist wanting to develop projects based on Arduino as its main microcontroller platform or an engineer interested in knowing what the Arduino platform offers, then this book is ideal for you.

If you have little or no previous experience in these kinds of tools, this book will help you get a complete view of the platform and the wide peripherals it has to offer by following a carefully designed set of project examples that cover the most important platform features.

Whether you have never written a line of code or you already know how to program in C, you will learn how to work with Arduino from the point of view of both hardware and software thanks to the easily understandable code that accompanies every project that has been developed exclusively with that premise in mind. This will be easy for those who don't have previous experience in programming.

This book was written with the aim to present the Arduino platform to all those wanting to work with Arduino but without any great knowledge of the microcontrollers scene. It will gradually develop a wide set of projects that have been designed to cover the most important aspects of the Arduino platform, from the use of digital and analog inputs and outputs to harnessing the power of interrupts.

What this book covers

Chapter 1, Meeting the Arduino Family, introduces you to the Arduino platform, and the different board models that integrate the Arduino family are presented, noting their common aspects.

Chapter 2, The Arduino Development Environment, shows you how to download, install, and set up a working Arduino integrated development environment and gives a complete explanation of its use and commands.

Chapter 3, Interacting with the Environment the Digital Way, covers the connection and use of digital outputs by dealing with simple devices that can be digitally operated, such as LEDs.

Chapter 4, Controlling Outputs Softly with Analog Outputs, shows you how to manage analog outputs and the use of transistor drivers to deal with high-current devices, such as motors.

Chapter 5, Sensing the Real World through Digital Inputs, explains the use of digital inputs by giving examples of typical applications, such as buttons and switches, and proposes an optical coin detection device that uses an optocoupler.

Chapter 6, Analog Inputs to Feel between All and Nothing, presents analog inputs and their use and offers two new projects: an ambient light meter with a photocell and a motor speed controller by using a potentiometer as an input device.

Chapter 7, Managing the Time Domain, introduces you to the different tools the Arduino library offers to deal precisely with time by building two more projects: a simple timer and a visual and acoustic metronome.

Chapter 8, Communicating with Others, shows you how to connect your Arduino projects to other platforms via serial communication and how to use the Serial Monitor to read from and send data to Arduino. A computer-controlled motor speed driver and a dial thermometer will be built.

Chapter 9, Dealing with Interrupts, shows you how to use interrupts to deal with unexpected events and to understand the difference between having to wait for something to occur and be called when it occurs. We will use a tachograph as a good example to show you all these concepts.

Chapter 10, Arduino in a Real Case – Greenhouse Control, gives you a complete real example of a project that summarizes all the concepts learned throughout the book.

What you need for this book

To work on all the projects shown throughout the book, you will need an Arduino board with its USB cable and a computer running Windows, Mac OS X, or Linux to program your board.

For the electronics circuits that will be built, a breadboard, some jumpers, and an assortment of the most common electronic components will be required.

The complete list of components used all along the different projects is as follows:

- A bunch of resistors
- Some LEDs
- Diodes and small transistors
- Switches and push buttons
- An optocoupler or optical switch
- A photocell
- A buzzer or small speaker
- Some potentiometers
- A thermistor
- A servomotor

Regarding previous knowledge, there is no need to know how to program because projects come with the entire code ready to run, and I will try throughout the book to introduce and clarify every programming aspect in the code.

Who this book is for

This book can be useful to a wide range of readers. It can be really illustrative to those wanting to be introduced to the development of projects based on microcontrollers and using Arduino in particular for the first time.

It can also be interesting to all those who already know or have worked with microcontrollers previously but haven't tried Arduino and still want to know the basics about this powerful platform by way of a number of projects that will present all important aspects of the platform.

Conventions

In this book, you will find a number of text styles that distinguish between different kinds of information. Here are some examples of these styles and an explanation of their meaning.

Code words in text, database table names, folder names, filenames, file extensions, pathnames, dummy URLs, user input, and Twitter handles are shown as follows: "Under Mac OS X, the installation of the application consists only of dragging the application icon to the `Applications` folder of your computer."

A block of code is set as follows:

```
void setup() {
  pinMode(transistorBase, OUTPUT);
  // Init serial communication
  Serial.begin(9600);
}
```

New terms and **important words** are shown in bold. Words that you see on the screen, for example, in menus or dialog boxes, appear in the text like this: "You'll have to go to **Control Panel** and locate **Device Manager**."

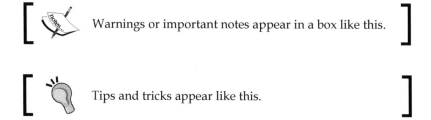

Warnings or important notes appear in a box like this.

Tips and tricks appear like this.

Reader feedback

Feedback from our readers is always welcome. Let us know what you think about this book—what you liked or disliked. Reader feedback is important for us as it helps us develop titles that you will really get the most out of.

To send us general feedback, simply e-mail `feedback@packtpub.com`, and mention the book's title in the subject of your message.

If there is a topic that you have expertise in and you are interested in either writing or contributing to a book, see our author guide at `www.packtpub.com/authors`.

Customer support

Now that you are the proud owner of a Packt book, we have a number of things to help you to get the most from your purchase.

Downloading the example code

You can download the example code files from your account at http://www.packtpub.com for all the Packt Publishing books you have purchased. If you purchased this book elsewhere, you can visit http://www.packtpub.com/support and register to have the files e-mailed directly to you.

Errata

Although we have taken every care to ensure the accuracy of our content, mistakes do happen. If you find a mistake in one of our books—maybe a mistake in the text or the code—we would be grateful if you could report this to us. By doing so, you can save other readers from frustration and help us improve subsequent versions of this book. If you find any errata, please report them by visiting http://www.packtpub.com/submit-errata, selecting your book, clicking on the **Errata Submission Form** link, and entering the details of your errata. Once your errata are verified, your submission will be accepted and the errata will be uploaded to our website or added to any list of existing errata under the Errata section of that title.

To view the previously submitted errata, go to https://www.packtpub.com/books/content/support and enter the name of the book in the search field. The required information will appear under the **Errata** section.

Piracy

Piracy of copyrighted material on the Internet is an ongoing problem across all media. At Packt, we take the protection of our copyright and licenses very seriously. If you come across any illegal copies of our works in any form on the Internet, please provide us with the location address or website name immediately so that we can pursue a remedy.

Please contact us at copyright@packtpub.com with a link to the suspected pirated material.

We appreciate your help in protecting our authors and our ability to bring you valuable content.

Questions

If you have a problem with any aspect of this book, you can contact us at questions@packtpub.com, and we will do our best to address the problem.

Disclaimer

Arduino brand, Arduino logo, design of the website, design of the boards, and all the board pictures used in the book are copyright of Arduino SA and cannot be used without formal permission. For information about the right way to use them, please write to trademark@arduino.cc.

All references in the book to Arduino should be considered as Arduino™.

1
Meeting the Arduino Family

In this first chapter, we are going to take a look at the microcontroller scene before the **Arduino** platform was presented. We'll see the changes it underwent that have made it a great success and that have led to it being widely adopted by hobbyists and technology lovers all around the world.

We will also meet some of the more popular Arduino models and compare their characteristics and features so that you can decide which model can be used in your next project.

A game changer

The introduction, in 2005, of the Arduino platform brought a totally new panorama to the microcontroller scene of the moment.

By that time, working with microcontrollers implied to pay quite a big price just for the microcontroller-integrated circuit itself and all necessary components and circuitry needed to make it work, and even to pay a much bigger price for the development tools needed to program it.

These development tools were rarely made publicly available and in most cases, they were mostly based on proprietary languages or, in the best of cases, in the assembly language, none of which were especially easy to learn for nonadvanced users. On the other hand, the user support was normally restricted and limited only to the manufacturer's microcontroller.

The Arduino platform changed every one of these aspects.

To begin with, it is an open hardware platform that is not only limited to the microcontroller integrated circuit, but it also provides a full board with all the necessary elements to power it, make it work, and connect it to a computer at a fraction of the price of most other microcontrollers available in the market at the moment.

On the other hand, the development environment was made freely available from the first moment as an open source project, consisting of a very simple and intuitive editor with its integrated compiler, based on a subset of the standard, well-known, and documented C language, widely available for many other platforms and architectures. We could say that the Arduino **Integrated Development Environment** (**IDE**) is just C with a friendly wrapper around it.

Last but not least, the Arduino online community was born: it allows thousands of users to share their ideas, projects, and philosophy, which makes them all support each other.

We are going to introduce some of the more popular Arduino models and compare their features and technical characteristics so that you can decide which model best suits the needs of your project.

Common features

The whole Arduino family, except Arduino Due, is based on 8-bit Atmel AVR microcontrollers, specifically those in the megaAVR series, ranging from the basic ATmega168 series used in the first Diecimila and Duemilanove models to the powerful ATmega2560 series of the latest Mega 2560 boards.

Apart from the ATmega series, the Arduino board incorporates every other electronic component necessary to make the microcontroller operate, including a crystal oscillator to set the working frequency that makes it run at up to 16 MHz.

Most models come with a USB port that enables you to connect the board directly to your computer so that you can program the microcontroller. The smaller models replace the USB connector with a direct RS-232 connection, which forces you to use a special USB-to-serial cable to connect the board to your computer.

Talking about powering, almost every model offers a power regulator that allows you to power the board directly from the USB port or through the jack connector, to power the board from an external power source such as a battery pack or a wall cube.

Finally, let's see a characteristic that also was an innovation when it was first introduced. The Arduino boards expose the microcontroller pins to the user through the use of two rows of female 0.1 inch headers, which makes it very convenient to connect the external components of your project, usually mounted on a breadboard, to your Arduino board with the simple use of jumper wires.

This distribution of external pin headers has made it possible to develop quite a bunch of different shields, which we will talk about later, that allow subassemblies to be easily connected to the main Arduino board.

A good number of models share the external footprint that has made the Arduino board so easily recognizable, including Arduino Diecimila, Duemilanove, Uno, Ethernet, and Yún.

For me as a programmer, one of the most important features is the fact that all members of the family share the same integrated development environment and language. This is the real common feature for all families.

Given the open hardware conception of the Arduino platform, the Arduino team has made publicly available the schematics, reference documentation, and even the EAGLE CAD files of all their boards.

For a full comparison of all Arduino boards' features and technical characteristics, you can visit the Arduino site at `http://arduino.cc/en/Products.Compare`.

Let's now take a closer look at the most popular board models and their specific features and configurations.

Arduino Uno

The Arduino Uno model is the evolution of the first Arduino board through the Arduino Diecimila and Duemilanove models.

A small footprint and a good pack of devices and available pins has made it the favorite board for beginners and advanced users who don't need great specifications to prototype and develop micro-controlled projects.

It is the most basic and cheapest model and is, in some way, the board that has made the Arduino platform what it is today.

Arduino Uno

It is based on the ATmega328 series, the descendant of the first ATmega168 series used in its older brothers, the Diecimila and the Duemilanove models. It works at a frequency of 16 MHz thanks to the use of a ceramic resonator, and offers a total of 32 KB of flash memory available to store your programs.

The Arduino Uno model exposes a total of 20 pins, 14 of which are digital input/output pins and 6 are analog inputs. Of the 14 digital pins, 6 can be used as analog output thanks to the included **Pulse Width Modulation** (**PWM**) mechanism, but we will talk in more detail about this in *Chapter 4, Controlling Outputs Softly with Analog Outputs*.

Some other peripherals offered by the board include serial and SPI communication ports, two external interrupts, an integrated LED (connected to pin **13**), and an external Reset button.

The board comes with a type B USB connector that makes it very convenient to connect to your computer or even power it from any USB output, including your own PC or any wall cube designed to serve as a USB charger.

As mentioned in the preceding section, it also comes with a jack connector that allows it to be powered from an external power source, such as a battery or an AC/DC adapter.

In any case, an input voltage between 7V and 12V is recommended to power the board, even though 6V to 20V can be accepted.

For a more detailed specification of its characteristics and available peripherals, you can visit the Arduino site and take a look at its product page at http://arduino.cc/en/Main/ArduinoBoardUno.

It is the perfect board to get introduced to the Arduino platform—available through a lot of different providers, some of which could be really close to you—or through the new Arduino Store at http://store.arduino.cc. Its price is around $25, which is quite a cheap price for the brain of your next project.

Arduino Mega 2560

Traditionally, the Arduino Mega 2560 model has been the offer the Arduino team made for those who need a more powerful board with a wider number of pins than Diecimila or Duemilanove.

Its footprint differs from that of Arduino Uno, making it a little longer than its little brother.

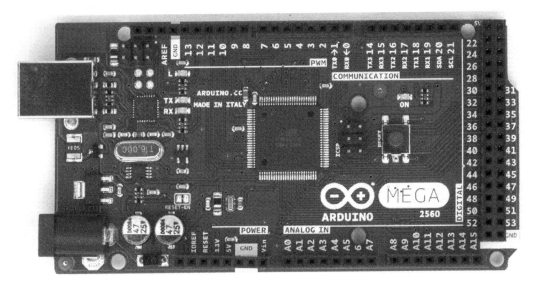

Arduino Mega 2560

The Mega 2560 model is based on the super powerful ATmega2560 microcontroller, just like Arduino Uno, which works at a clock frequency of 16 MHz. However, unlike its little brother, it comes equipped with a vast number of peripherals.

One of the biggest advantages of the Mega 2560 model is the size of its flash memory, which goes up to 256 KB, equivalent to eight times the memory space offered by Arduino Uno, making the Arduino Mega 2560 model the target platform for those projects that need complex software to manage them.

On the other side, it has no less than 54 digital input/output pins, of which you can use 15 for analog output through the use of PWM and 16 analog input pins.

Related to its other characteristics, the Mega model comes with four serial communication ports, an SPI communication port, and a total of six different hardware interrupts.

Just like the Uno model, the Mega model implements one integrated LED and external Reset button.

When talking about powering the board, the Mega model can be powered just like the Uno model, that is, through the USB connector or through the external jack, supporting the same input voltage ranges as the Uno model (7V-12V recommended but 6V-20V accepted).

You can find its detailed specifications, schematics, and some more reference documentation about its product page on the Arduino site at `http://arduino.cc/en/Main/ArduinoBoardMega2560`.

This board is ideal for those who have to deal with projects that have wider requirements, in particular, those related to the number of input/output pins to interact with external devices and to the flash memory size needed to store the programs, allowing for much more complex programs than the Arduino Uno model.

Arduino Ethernet

On the lines of the Internet of Things, the Arduino team presented the Arduino Ethernet board, which included an Ethernet interface, making it able to develop projects that were connected to an Ethernet network or to the Internet itself.

Arduino Ethernet

The board is developed around the same ATmega328 microcontroller that is present in the Arduino Uno model, so the memory size and other characteristics are the same as in the Uno model.

The inclusion of the Ethernet module, however, forced some restrictions on the Arduino Ethernet model.

On one hand—and although the number of digital input/output pins is still 14 just like in the Uno board—some of them are reserved to interface with the Ethernet module, allowing only the use of nine of them for your project. On the other hand, the Arduino Ethernet board doesn't include a USB port. So, you have to use a dedicated six-male pin connector to upload your programs to the microcontroller through the use of a special USB-to-serial converter, usually known as an FTDI cable, due to the use of the FT232RQ chip around which the converter is normally built.

The Arduino Ethernet model comes with an integrated Micro SD card reader that allows you to store files and resources of a bigger size that are going to be served over the network. To get access to the SD card, the use of an external library is needed.

The full characteristics list of the Arduino Ethernet is on its product page on the Arduino site at `http://arduino.cc/en/Main/ArduinoBoardEthernet`.

Given that there is no USB port on the Arduino Ethernet board, the possibilities for powering the board are a little different. You can power your board through the external jack from an external power source by the use of the previously mentioned FTDI cable from a USB port or by the additional **Power over Ethernet (PoE)** module, which allows the board to draw current from the Ethernet connection itself, requiring your project to only be connected to the Ethernet network in order to start working.

The PoE module is not available as an add-on or as a shield, but it has to be ordered with your Arduino Ethernet board when you buy it. You can take a closer look at its product page in the Arduino Store at `http://store.arduino.cc/product/A000061`.

LilyPad Arduino

Apart from the previously mentioned Internet of Things, currently there is another line of product development called wearables, consisting of products you can wear as part of your clothing and that constantly interact with you and your environment as if they were an enhancement of your own body.

In this line of wearable technologies, the Arduino family incorporated a design by Leah Buechley, the LilyPad Arduino board, which is a little, round microcontroller device aimed at projects where it could be sewn to textile with conductive thread acting like normal wires.

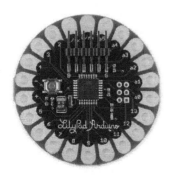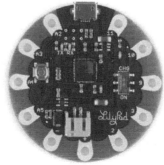

Arduino LilyPad subfamily

The LilyPad Arduino board is a little subfamily of boards by itself with four different models and specifications:

- LilyPad Arduino
- LilyPad Arduino USB
- LilyPad Arduino Simple
- LilyPad Arduino SimpleSnap

All of them come with different microcontrollers, ranging from the ATmega168V series to the new Atmega32u4 series with flash memory sizes between 16 KB and 32 KB.

They all are able to work under low voltages to reduce power consumption, and that is why they all work at a clock frequency of 8 MHz, which is half the frequency of its bigger brothers.

One thing to consider when choosing one of these models for your project is that none of them come with a serial communication port in case you need it.

The small dimension of the LilyPad Arduino board (approximately 50 mm in diameter) considerably reduces the number of pins and features it can offer. Only the LilyPad Arduino board has the same specifications as the Arduino Uno board with respect to the number of available pins, that is, 14 digital input/output pins with six PWM and six analog inputs. The rest of the LilyPad boards offer only nine digital input/output pins, of which four can be used as the analog output with PWM and four others that can be used as analog inputs.

In the case of the LilyPad subfamily, the available pins are not offered via a female header as is the case with other Arduino boards; in this case, they offer a number of silver-plated holes that allow you to sew the conductive thread to them. In the case of the LilyPad Arduino SimpleSnap board, the holes are replaced by snaps that easily allow you to connect and disconnect the external devices to and from the main Arduino board.

With respect to the connection to your computer, they also have to restrict the USB port and change it for a six-male pin header, having to connect the board to the computer through the FTDI cable mentioned previously. Only the LilyPad Arduino USB offers a micro USB port.

As a unique case, the LilyPad Arduino SimpleSnap board also incorporates a lithium polymer battery and all the circuitry required to charge it.

All of them can be hand washed except for the LilyPad Arduino SimpleSnap board, due to its battery, but in this case, disconnecting it from the textile should be easy, thanks to the integrated snaps.

You can find a complete list of features of every model on its own product page on the Arduino site at `http://arduino.cc/en/Main/ArduinoBoardLilyPad` for the LilyPad board, `http://arduino.cc/en/Main/ArduinoBoardLilyPadUSB` for the LilyPad USB board, `http://arduino.cc/en/Main/ArduinoBoardLilyPadSimple` for the LilyPad Simple board, and `http://arduino.cc/en/Main/ArduinoLilyPadSimpleSnap` for the LilyPad SimpleSnap board.

In definite terms, the small dimensions and reduced power consumption of this Arduino board make it ideal to develop wearable projects, where they can be directly integrated with clothes and are going to be running for long periods of time.

Arduino Yún

One of the latest incorporations to the family has been the Arduino Yún board, bringing a slightly different approach to the concept of the Internet of Things.

In this case, the board incorporates two different interconnected sections. On one hand, it has a usual Arduino board with its external pins and all the devices that we have already seen, but on the other hand, it is a totally operative Linux device with Ethernet and Wi-Fi support.

This way, you can build projects that benefit from both parts: the Arduino ease of use to interact with the real world and the power and connectivity options of a Linux device in which you can develop your own shell scripts or small programs with languages such as Python.

Arduino Yún

Taking a more detailed look at the technical specifications of this new board, we can realize the power we get to develop real Internet of Things projects.

With respect to the Arduino side, it comes equipped with an ATmega32u4 board running at 16 MHz, exposing 20 digital input/output pins (seven of which can allow PWM to be used as analog outputs) and no less than 12 analog inputs.

On the Linux side, it runs on an Atheros AR9331 chip at a clock frequency of 400 MHz with an internal flash memory that acts as its main hard disk with a total size of 16 MB of which around 9 MB is dedicated to store OpenWrt, the Linux operating system inside the Yún. If you need more space to store resources needed by your projects, Arduino Yún is also equipped with a micro SD card slot that you can use as additional storage.

Meeting the Arduino Family

The whole device can be powered through a micro USB connector, but not having a voltage regulator makes it difficult, but not impossible, to power it from an external battery pack.

You can visit its product page so that you can get a full picture of this model board at http://arduino.cc/en/Main/ArduinoBoardYun.

The Arduino Yún board is, no doubt, a perfect device to develop not only physical projects, but also to connect them to the network.

Arduino Mini, Micro, and Nano

Another very popular subfamily inside the bigger Arduino family is the one formed by the Arduino Micro, the Arduino Mini, and the Arduino Nano boards; the following screenshot follows the same order as the one mentioned:

Arduino Micro, Mini, and Nano

These little boards, not much bigger than a postal stamp (approximately 2 inch by 0.75 inch), get, in most cases, very similar characteristics to that of their bigger brothers. They are just packaged in such a way that makes them ideal to connect directly to the breadboard or in projects where space is a must and the controller has to be small enough to be embedded in the general assembly.

To be more specific with its technical specifications, the Arduino Micro board comes with an ATmega32u4 microcontroller and both the Mini and Nano boards come with an ATmega328 chip; all of them run at a clock frequency of 16 MHz and come equipped with a flash memory of 32 KB to store your sketches.

With respect to the available pins, both the Mini and the Nano boards offer just the same number of pins as the Uno board but with two more analog inputs, making a total of eight. The Arduino Micro board goes even further with 20 digital input/output pins, seven of which can use PWM to act as analog outputs, and a total of 12 analog inputs.

Talking about how to power the boards, the Arduino Mini board simply doesn't have any facility, as it's your responsibility to provide a regulated power supply to allow the board to run. Both the Arduino Micro and Arduino Nano boards come with an integrated USB port with mini and micro USB connectors that you can use to power the board or even the final project from a typical USB phone charger.

Finally, except for the Mini board, the Micro and Nano boards come with a serial port that you can use to communicate your project with other serial capable devices.

They all share the same philosophy: to develop just the core of the microcontroller so that the size of the board can be reduced at a maximum, allowing for smaller boards suitable to be incorporated in your projects without using as much space as other bigger boards of the family.

Other Arduino family members

If all the boards shown at the moment don't seem enough to you or you still haven't found the model that best suits the needs of your next project, don't worry. The Arduino team has also developed some other models with very different characteristics and orientations, but for the purpose of this book, they are all similar to the models already shown. They share the same general philosophy and, perhaps more importantly, can be programmed in just the same way with just the same tool.

In any case, we will just show them briefly without entering into too much detail and just revealing some of their more remarkable features and uses.

Esplora

The Esplora board is the all-in-one board of the family. The main difference between this board and the rest of its brothers is that this board is not only a microcontroller board, but also a good selection of different sensors and input devices, such as a temperature sensor, accelerometer, microphone, or a joystick and some outputs, such as sound and light.

It is designed for those who have little or no previous electronic knowledge but want to start working with the Arduino platform from the very first moment.

Arduino Esplora

The Esplora product page contains its full characteristics' list and information at http://arduino.cc/en/Main/ArduinoBoardEsplora.

Arduino Robot

Robotics is one of the disciplines that is constantly pushing forward and demanding for richer and more powerful advances in the microcontrollers arena, because as the microcontroller has more features, it is easier to build a robot with it.

As it is a very attractive discipline for kids and students, it is the perfect introduction to electronics and their microcontrollers.

This is why the Arduino team developed the Arduino Robot model, a fully featured, totally operable robot that shares the microcontroller it is based on with the other boards and, of course, the language it will be programmed in and the tool we will use to program it.

Arduino Robot

To be precise, the Arduino Robot model is really composed of two interconnected boards, a motor and a control board, each with its own microcontroller and dedicated to very different tasks.

Among others, it is equipped with LCD, compass, speaker, a bunch of LEDs, five buttons in the control board, IR sensors, motor drivers, motors, and wheels in the motor board.

You might be interested in taking a look at the Arduino Robot product page at `http://arduino.cc/en/Main/Robot`.

As discussed, Arduino Robot is a totally operative robot that you can program just the same way you do with your other Arduino boards by squeezing a little more off your budget.

Arduino Due

The last board that I will show you is the Arduino Due board, which in some way, is the evolution of the Arduino Mega board but with an Atmel SAM3X8E ARM Cortex-M3 CPU, making it the only family member capable of 32-bit operation.

It runs at a clock frequency of 84 MHz and has an astonishing list of features among which the standout is its 512 KB of flash memory, 54 digital input/output pins, or four serial ports.

Another remarkable characteristic of the Due board is that it operates at 3.3V, being the only member of the family that operates at this voltage. From the point of view of powering the board, you shouldn't care because it comes with a voltage regulator that allows you to power the board from the USB port from an external battery or power source, but it can be a serious problem if you plan to use a shield designed for another member of the family that operates at 5V.

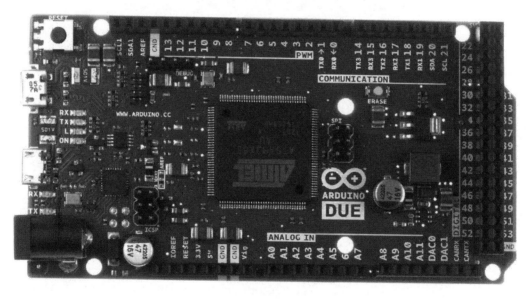

Arduino Due

Unofficial boards

Given the open hardware nature of the Arduino project, there are now a lot of different boards. All of them are compatible with the original Arduino board, and some of them even extend the characteristics of its predecessor.

In some cases, they can even be found at just a fraction of the price of the original Arduino board, and, sadly, also at a fraction of the quality in other cases, so be especially careful when choosing your provider.

For a complete list of some of these Arduino clones, you can take a look at this list at Wikipedia: `http://en.wikipedia.org/wiki/List_of_Arduino_boards_and_compatible_systems`

Shields

There is a group of boards that, not being Arduino boards, are directly related to Arduino. They are called shields and are small add-ons that you can directly plug onto your Arduino board and that have all the necessary electronic components and circuitry to accomplish the mission they have been designed for.

There are hundreds of shields for Arduino out there. Just perform a simple search on Google and the number of results will amaze you: the results range from a GPS shield that allows our project to be location-aware to shields that link two Arduino boards via radiofrequency, passing by GSM/GPRS shields that make our project able to establish radio connections.

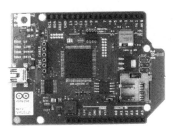  

Some official Arduino shields

Furthermore, when most users finish the development of a project, they opt to create this shield to transfer the complementary components out of the breadboard.

The Arduino team has developed some official shields for Arduino, among which there is the Arduino GSM Shield that allows GPRS communications, the Arduino Wi-Fi shield that enables Wi-Fi connections, or the Arduino Motor shield that eases the operation of external motors.

The Arduino platform has grown so much that today, there is a whole market around it, with providers developing shields for unimaginable purposes.

Among the biggest shield developers are Adafruit (http://www.adafruit.com/category/17_21), SparkFun (https://www.sparkfun.com/categories/240?page=all), and Cooking Hacks (http://www.cooking-hacks.com/shop/™Arduino/shields).

Just one to rule them all

I have already shown you all the significant Arduino models you might need to develop your project. As they are so different from each other, they all share a very important characteristic. They can all be programmed using the same language and the same tool.

This is a wonderful thing, because it allows you to program all models once you know how to program just one, making the selection of the board your unique concern.

In the next chapter, we will prepare our developing environments so that we can directly begin to program our boards and test our first assemblies.

Users teaching users

I told you previously that Arduino was not just a board and a compiler. One of the things that I personally think has made Arduino become such a big platform is its online community.

On the Arduino site, you can find not only a traditional forum (http://forum.arduino.cc) where lots of users share their projects and troubles and help others, but also one of the sections I like the most: the Arduino **Playground** section.

The Arduino **Playground** section (http://playground.arduino.cc) is a section of the Arduino website where users publish information about ways to connect Arduino to other devices of all kinds, linking to their own websites in most cases filled with information about the project they have worked on. It is some kind of collaborative showroom.

For me, the **Playground** section is, perhaps, the most valuable site you can go to whenever you are beginning to work on your own project. There are plenty of possibilities that you can find in that piece of information that can serve you as a starting point.

Summary

In this chapter, we took a look at the microcontroller context before and after the introduction of the Arduino platform. We met some of the family members and we came to know about their most significant features so that you can decide which one of them best matches the requirements of the project that you are thinking about.

We also learned about the Arduino extensions, called shields, and saw some of the official shields developed by the Arduino team.

In the next chapter, we will download, install, and take a tour of the Arduino IDE, and we will finish by uploading our first sketch to our board, so roll up your sleeves as we begin to enjoy working with our favorite microcontroller.

2
The Arduino Development Environment

In this chapter, I'll show you how to get up and running with the Arduino development environment. We will download and install it, and we will take a tour through all its menus and commands. We will finish by uploading a first sketch to your board so that you can confirm that all that is needed to begin is working correctly.

A multiplatform tool

One of the things I like the most about the Arduino software is that it is truly multiplatform, which means that it is exactly the same environment whether you run it under OS X, Windows, or Linux. You could find some differences in the installation process of every operating system and differences to get it up and running, but once you have it up and running, it is just the same in any platform.

So, let's go ahead and present to you the whole process to make it run under the operating system of your choice.

Downloading the package

The first thing you have to do is go to the Arduino site's download section at http://Arduino.cc/en/Main/Software and choose the right package for your operating system and the software branch that best suits your board.

At the time of writing this, there are two different branches. I'll recommend that you use the Arduino 1.0.X stable branch unless you are going to work with Arduino Yún or Due; in that case, you'll have to choose the 1.5.X branch that, at the time of writing this, is a beta version and, therefore, could be a little unstable or could lead to some errors. However, I have to say that I have been working for more than a year with the 1.5.X branch to program both Arduino Uno and Arduino Yún, and I haven't seen any bugs.

On the Arduino site, you will find precompiled packages for Mac OS X, Windows, and Linux and even a source code package.

Let's take a closer look at each of the supported operating systems.

Windows

For Windows, there are two different precompiled packages: one prepared for the Windows Installer and another that consists only of a ZIP package that you have to uncompress by yourself.

The Windows Installer option is recommended in order to allow multiple users in your computer to access the software, but you need administrator privileges to install it system-wide.

If you don't have an administrative account, you can download the ZIP compressed package and simply uncompress it in any folder where you have permission to do so, such as your own desktop.

Mac OS X

For Mac OS X, there is little to consider. There is just one package so that you have nothing to choose from. Just download your package and go to the next section to see how to install it.

Linux

For Linux, the only consideration to have in mind is the architecture of your computer so that you have to choose between the 32- and 64-bit precompiled packages.

In both cases, the package provided in the Arduino site consists only of a compressed TGZ file that you have to uncompress by yourself.

Source code

If you use a different operating system, there is even a source code package that allows you to compile the Arduino development environment on your own.

Installing the software

We have seen the downloading considerations for every operating system; let's now go on to see how to install the package for each of them.

Windows

Under Windows, using the Windows Installer package is quite simple and doesn't require any special consideration.

If you opted for the compressed package, you have to only uncompress it with an archive uncompressor such as WinZIP to make it available, which is not too hard.

Mac OS X

Under Mac OS X, the installation of the application consists only of dragging the application icon to the `Applications` folder of your computer. Simple.

In the latest versions of Mac OS X, Java may not be preinstalled; if this is the case, you should go to the Java official website at `http://www.java.com/en/download/` and download it.

Linux

Given the diversity of different distributions and the packages' dependencies system on which Linux relies, you should install some of them before you can run the Arduino development environment.

Thanks to the package management systems based on centralized repositories present in most Linux distributions nowadays, it is even possible to install the Arduino development environment directly from a repository along with all the needed dependencies using just a command. It makes this the preferred way of installation if the available version in the repository is up to date.

> In Debian-based distributions, the Arduino development environment could be installed as easily as running the following command:
> `apt-get install arduino`

In case of trouble

In case you encounter any problem during the installation process, you could go to the Arduino site and take a look at your corresponding operating system's **Getting Started** guide for more detailed step-by-step instructions at http://Arduino.cc/en/Guide/HomePage.

Installing the drivers

As it is a USB device, the Arduino board might need its own drivers to be installed before your computer can talk to it.

The most recent boards, such as the Arduino Uno or the Arduino Mega 2560 board, don't need any special drivers to be installed. If you are using an older model, such as the Diecimilia board or any other board with an FTDI driver chip like the one shown in the following screenshot, you will have to install the specific drivers:

Chapter 2

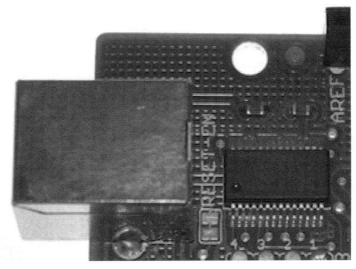

FTDI Chip

You can download the driver's version for your operating system right from the FTDI manufacturer page at http://www.ftdichip.com/Drivers/VCP.htm.

We will provide brief instructions on how to install these drivers for the operating system of your choice.

Windows

Under Windows, the installation may be a little tricky due to the fact that Windows will do its best to recognize the board and locate appropriate drivers for it, failing most of the times in doing so. To get the drivers up and running, perform the following steps:

1. You'll have to go to **Control Panel** and locate **Device Manager**.
2. Once you're there, look for a device with an exclamation sign, usually under the **Ports** section (also, take a look at the **Other Devices** section if you don't find it under **Ports**).
3. Once you have found it, right-click on it, select the **Update Driver Software...** option, and navigate to the folder on your hard disk where you uncompressed the FTDI drivers you downloaded from the previously mentioned URL of the FTDI site.

If you still can't get it correctly installed, take a look at the **Getting Started with Arduino on Windows** section on the Arduino site at `http://Arduino.cc/en/Guide/Windows`.

If you are running Windows XP, there is even a step-by-step guide with screenshots of every step on how to install the Arduino Uno driver under this operating system at `http://Arduino.cc/en/Guide/UnoDriversWindowsXP`.

Mac OS X

As stated previously, drivers' installation is only needed if you are using an old board, such as the Diecimila or Duemilanove board. In this case, you'll have to download them from the previously mentioned URL, double-click on the uncompressed package installer to install it, and reboot your computer after the installation.

Linux

The Linux kernel supports the FTDI drivers without the need to install any additional software, but if you are experiencing any trouble, you can take a look at the **Installing Arduino on Linux** section of the Arduino **Playground** tab at `http://playground.Arduino.cc/Learning/Linux`, where you can find more detailed information specific to the more common Linux distributions nowadays and, in particular, to the libraries and other' requirements your installation should meet.

Running the Arduino development environment for the first time

Well, it may have seemed more complicated than it really was, but finally, you have your programming environment ready to work. It's time to create our first test, take a tour of it, and meet all its parts and structures.

On the first run, the Arduino development environment should look like this:

Arduino development environment

Basically, it's divided into three sections:

- The toolbar at the top with buttons for the more usual commands
- The code editor in the middle, where you will write your sketch, as we commonly call an Arduino program's code
- The message area at the bottom, where you will get information about the status of your sketch and possible location of errors in it

Let's take a closer look at each one of these zones so that you can begin to use them.

The toolbar

In the toolbar, you're going to find a total of six buttons to call the more usual commands of the development environment.

Buttons on the toolbar

From the left-hand side to the right-hand side, they are as follows:

- **Verify**: This button is used to verify the syntax in your code and compile it if no errors are found.
- **Upload**: You can click on this button to upload the resulting machine code to the Arduino microcontroller. It will compile the code if it has not been compiled previously.
- **New**: This opens a new blank sketch.
- **Open...**: This loads a previously saved sketch from the disk through the use of a pop-up menu.
- **Save**: This saves the currently edited sketch to the disk.
- **Serial Monitor**: This opens the **Serial Monitor** window that allows you to visualize the interchange of data in a serial communication between your computer and the Arduino board. We will learn more about this in *Chapter 8, Communicating with Others*.

The code editor

The code editor is the zone where most of your work will happen and where you will spend most of your time.

It is a multitab editor, which means that you can open more than one document at the same time on different tabs. You can find a small downward-pointing arrow icon on the right-hand side of the tabs area that unfolds a menu with different tab-related options, such as **New** and **Rename**, or helps you move along the currently opened tabs.

It is not a very powerful editor but has all the features that one can hope to find in a modern code editor nowadays, such as syntax highlighting, which means that the different parts and words of your code are going to be presented in different colors that will help you classify the different elements in it, making it clear what a reserved word, a variable, or a constant is.

Another feature I really appreciate and that makes your work much more comfortable is the highlighting of parentheses and bracket pairs when you put the cursor over one of them so that you can easily view the opening/closing pair of such constructions.

The code editor showing syntax highlighting and the pop-up menu

These features apart, there are some other useful commands, such as **Search in Reference**, that take the word under the cursor and look for it in the **Arduino Language Reference** section on the Arduino website, showing you the help page of the reserved word or function.

There are other stylistic commands, such as **Increase Indent**, **Decrease Indent**, and **Comment/Uncomment**, which are available through the pop-up menu that is shown when you right-click on the editor area.

The message area

The message area is the zone where the Arduino development environment will tell you which errors it has found when trying to compile your code or any other issues it has found when trying to accomplish the command you asked for, for example, a communication error when uploading the machine code to the microcontroller board.

The message area showing a syntax error and line highlighted in the code editor

When the error is related to a line in your code, this will also be highlighted in the code editor, as shown in the previous screenshot.

The Arduino development environment will try its best to reference in the message area the line number where it thinks the error is. I say *thinks*, because the mistake is not always correctly located and, in some cases, it can be one or two lines before the line the message area is reporting. This is a typical situation when programing in C, but we will talk about it later on.

Preflight checks

In order to upload your first sketch to the Arduino board, you have to first make sure that the Arduino development environment knows two very important things about your board:

- The type of Arduino board you have
- The serial port through which it is connected to your computer

Both parameters have to be specified using the **Tools** menu in the menu bar and by selecting the **Board** and **Serial Port** commands.

In the next screenshot, you can see all available options when selecting the board type:

All board models available through the Board command in the Tools menu

Being a teacher myself and having worked with Arduino boards with my students for some years, I have found that the most common mistake they make the first time they try to upload their first sketch to the Arduino board is the wrong selection of the serial port. If you are like them and don't know for certain which one of the available serial ports in the **Serial Port** command to select, don't worry. In the worst case, it's just a matter of trying them all.

When the Arduino development environment can't communicate with your board due to an incorrect serial port specification, it will show you a message similar to the one you can see here:

```
Problem uploading to board. See http://www.arduino.cc/en/Guide/Troubleshooting#upload for suggestions.
Binary sketch size: 1,084 bytes (of a 32,256 byte maximum)
avrdude: stk500_recv(): programmer is not responding

1                                                           Arduino Uno on /dev/tty.usbmodem1411
```

A typical message shown when an incorrect serial port is selected

Uploading our first sketch

Once you know how to connect the board to your computer and how to let the Arduino development environment know about it, the moment has come to create our first real test.

We will upload the simplest sketch to the Arduino board just to confirm that all parts are correctly set up and that you can, for sure, begin to work with it and learn how to program it.

What we will do is load one of the examples that comes with the Arduino development environment, in particular, one called **Blink** that makes use of the onboard integrated LED to make it do just that—blink.

To do this, go to the **File** menu in the menu bar and select the **Examples** command, navigate to **01.Basics**, and select **Blink**.

The Arduino development environment should open a new window containing the following code for that example:

```
/*
  Blink
  Turns on an LED for one second, then off for one second,
repeatedly.

  This example code is in the public domain.
 */

// Pin 13 has an LED connected on most Arduino boards.
```

```
// give it a name:
int led = 13;

// the setup routine runs once when you press reset:
void setup() {
  // initialize the digital pin as an output.
  pinMode(led, OUTPUT);
}

// the loop routine runs over and over again forever:
void loop() {
  digitalWrite(led, HIGH);   // turn the LED on (HIGH is the
voltage level)
  delay(1000);               // wait for a second
  digitalWrite(led, LOW);    // turn the LED off by making the
voltage LOW
  delay(1000);               // wait for a second
}
```

>
> **Downloading the example code**
> You can download the example code files from your account at http://www.packtpub.com for all the Packt Publishing books you have purchased. If you purchased this book elsewhere, you can visit http://www.packtpub.com/support and register to have the files e-mailed directly to you.

We will go on with more details about the programming language and its characteristics in the coming chapters, but at the moment you just need to know that, it simply configures the Arduino board to use the onboard LED as an output in the setup function and turns it on and off infinitely, with a little delay in between, in the loop function, that is, blink an LED.

Once you have loaded your sketch, you only have to click on the **Upload** button on the toolbar (the one with the rightward-pointing arrow), which will cause the Arduino development environment to first compile the sketch if it wasn't already compiled and, if it finds no errors, it will upload it to the selected board through the specified serial port.

During the uploading process, you can see two other integrated LEDs labeled as **TX/RX** blinking, indicating that the serial communication has been established and that the Arduino board is receiving data from the computer and sending back an acknowledgment to it.

As soon as the uploading process has finished, the Arduino board will immediately begin to run the sketch and as a result, you should see the LED labeled **L** begin to blink at a 1 Hz frequency.

Congratulations, you have completed your first development cycle: edit the sketch (load it from the disk, in this case), compile it, upload it to the board, and run it there.

If that wasn't enough for you, perhaps you could try to simply change the two values in the `delay` functions from `1000` to, say, `300` and `500` and reupload it to the board to watch how the blinking frequency changes.

Main menus and commands

Before going on to the next chapter where we begin to deal with all the features present in the Arduino board, let's take a little tour of the Arduino development environment menus and commands.

It has only five menus, and some of them are very similar to those present on any other application in your computer, such as **File**, **Edit**, or **Help**.

In the **File** menu, you will find commands to open a new sketch, load it from the disk, save it, or print it. Just as an addition, you can see two commands: **Sketchbook** and **Examples**.

The **Sketchbook** submenu will unfold another submenu listing all sketches available in the folder you can designate in the Arduino development environment preferences as your sketchbook folder. This makes it very easy to load your own sketches without the hassle of having to navigate through your entire filesystem folder structure to find the particular sketch you are looking for. The only drawback of this is that you have to store your sketches in the specified folder.

On the other hand, the **Examples** command will do something similar by showing all examples that come with the environment, which is very convenient when learning to program.

In the **Edit** menu, you'll find options such as **Undo**, **Redo**, **Copy**, **Cut**, and **Paste** as with every other application in your computer, but you'll also find commands such as **Copy as HTML** to publish the code in a web page or **Copy for forum** to prepare the code in a convenient way to be published in the Arduino Forum.

There are also commands to embellish and tidy up your code, such as the previously mentioned **Increase Indent / Decrease Indent** and **Comment/Uncomment** commands, and, of course, commands to find, and to find and replace, like in most editors.

The **Sketch** menu has just four commands:

- **Verify/Compile**: This is just the same as the **Verify** button in the toolbar.
- **Show Sketch Folder**: This opens the folder in your disk that stores the sketch being currently edited.
- **Add File...**: This allows you to include one more file in the current sketch, allowing you to make your code modular by dividing it into different files.
- **Import Library...**: This shows you another submenu listing all available libraries in the `libraries` folder inside the designated sketchbook folder of your hard disk. We will talk in more detail about libraries in other chapters of the book.

In addition to the previously mentioned **Board** and **Serial Port** commands, the **Tools** menu has commands to format your code for better viewing or to create a ZIP package of your sketch to make it easier to send it via e-mail.

The **Serial Monitor** command is just the same as its corresponding button on the toolbar and opens the **Serial Monitor** window to allow us to communicate serially with the board and see what the board is sending us back.

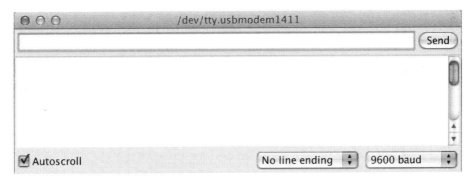

The Serial Monitor window

The **Programmer** and **Burn Bootloader** options are for advanced users who want to reprogram their Arduino boards' bootloader (a sort of operating system that makes the Arduino board run).

Finally, the **Help** menu contains commands to show different sections and views of the Arduino documentation.

The Arduino language

As you may have noticed, the code you write for your Arduino sketches is very similar to the C programming language. It is, in fact, a subset of it, which makes it very convenient to use, given that the C programming language is almost a de facto standard in most platforms and devices.

Traditionally, the C programming language has been portrayed as not being very user friendly, but I prefer to refer to it as being friend-selective.

One of the missions of this book is to introduce you to the C programming language, so that if you don't have any previous knowledge of it, you can take your first steps in an easy and secure way and ensure you will become good friends with it.

The only thing you should do is strictly respect its syntax and know that in the C language, a comma or a semicolon are important and that the case when writing is also significant. Perhaps these two things are the origin of half the mistakes you will make when programming and that give the C programming language its bad reputation.

Summary

This has been a long and, in some ways, an arid chapter, but it's totally necessary to prepare your development environment and to allow you to directly jump into the real assembly and programming with Arduino.

We installed both the software and the corresponding drivers, and we created the first test for the development environment and the board itself.

In the next chapter, we will deal with the real thing and begin to work with digital outputs and their corresponding code.

The show is ready to begin.

3
Interacting with the Environment the Digital Way

Now it's time to start working hands on with our Arduino boards by connecting electronic circuits to them and programming the microcontrollers to interact with the external circuitry.

In this chapter, we are going to deal with digital outputs, as it is the simplest way to interact with the outside world, and we will learn about the necessary code to drive them.

Let's begin with some considerations regarding digital and analog signals, move on to the circuit assembly, and finish by writing some code to make the microcontroller command the external circuit.

Digital versus analog signals

When working with electronics signals, you will constantly see references to digital and analog signals, and it is important that you differentiate between the two and know how to make Arduino deal with every type of signal.

A digital signal is one that takes only two clearly different states—no more, no less. To give you an example, switching a light on and off can be a typical case of a digital signal. You just have these two states, and it is on or it is off; it can't be both at the same time and nor can it have any other possible state in between.

In opposition to digital signals, analog ones are those that have a theoretically infinite number of possible values between a minimum and maximum one. Looking for an example relative to light, such as the one given for digital signals, we can think of the light coming from the sun through a window. It has a minimum, when there is no sun in the night, and a maximum, when the sun is just in front of your window. However, between these two extreme values, there are an infinite number of them, differentiating one another in a very small increment as long as the sun goes from the minimum position to its maximum.

When working with computers and microcontrollers, digital signals are often represented by the 0 and 1 value, 0 being the value used for the off state and 1 for the on state, which makes them perfect to be represented as binary digits.

In the Arduino language, we even have two more convenient constants to reference these two states, HIGH and LOW, as you may remember from the **Blink** code example we saw in the previous chapter. In any case, we will see more about this in this chapter.

On the other hand, when dealing with analog signals from the point of view of a computer, we usually represent them as real numbers, or floating-point numbers in computer science jargon, but we will talk in more detail about this in *Chapter 4, Controlling Outputs Softly with Analog Outputs*.

Our first circuit

To begin working with digital outputs, we are going to connect a very simple circuit to our Arduino board and write some code to deal with it.

In the rest of the book, I'll present you with different circuits that we will have to assemble and connect to the Arduino board. Regarding the external circuits assembly, nowadays we don't really have to solder all parts to a printed circuit board to get our circuit up and running. We can simply use a bunch of short wires, called jumpers, and a breadboard.

Before going on with our first circuit, let's take a look in detail at what a breadboard is and how to use it.

Using a breadboard

A breadboard is a square panel built in such a way that it allows the connection of the electronic component plugged into it without the need to solder them together or use any other form of connection among them.

A typical half size breadboard

As you can see in the preceding picture, the breadboard has a lot of holes distributed in four different areas:

- Two horizontal rails up and down with two rows of holes
- Two blocks for components' connections in the middle, usually with columns of five holes

You plug the components' legs into the holes. They are equipped with metallic clips inside that fit the electronic components once inserted. They are distributed in the breadboard in such a way that it connects all holes in every rail's row and every hole in every five-hole columns in the components' area.

To allow you to connect the circuit to the Arduino board or other external components not placed in the breadboard, you should use small pieces of wire that go from a hole in the breadboard to the external component leg.

For more detailed information, with very illustrative images and examples of use, you can visit the excellent Sparkfun tutorial on how to use a breadboard at https://learn.sparkfun.com/tutorials/how-to-use-a-breadboard.

At the moment, you don't have to worry about being very capable with a breadboard, because I'll give you detailed schematics and diagrams on how you should connect every one of the proposed circuits to your breadboard.

The LED circuit

The first circuit we are going to assemble is a simple LED with its current limiting resistance, and we will connect it to a digital output of our Arduino board.

Arduino has two different rows of pin headers:

- One with the powering pins and analog inputs in the lower side of the board
- The other in the upper side, with all digital pins

Take a look at the following screenshot:

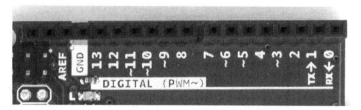

An Arduino row of digital pin headers

You can see that some of the pins in the digital row are marked with ~, which, as stated in the board serigraphy, are available for use as analog outputs through the use of **Pulse Width Modulation (PWM)**, about which we will talk more in the next chapter. This doesn't mean that you can't use them as digital pins, but it simply means that when you are going to use analog outputs, you should use only those pins that are marked **PWM**.

For our example, we will connect a current limiting resistance of about 220 Ohms to the Arduino pin **12**, the other leg of the resistance to the LED anode, and finally, the LED cathode will be connected to the ground of our Arduino board, available at the bottom pin header through any of the two pins marked as **GND**.

An Arduino row of power and analog pin headers

This way, we will power the LED directly from the Arduino digital output pin in a programmatic form, and it will be our code that will determine when to turn the LED on and off by just setting that digital output pin HIGH or LOW, which will set it to 5V or 0V, respectively, thus providing or not providing current to the LED.

Circuit schematic

I'll always try to give you an electronic schematic of our assemblies so that you can understand what we are going to connect when using the breadboard or if you prefer any other method to make the necessary connections.

The complete assembly for the connection of the LED is as shown in the following schematic:

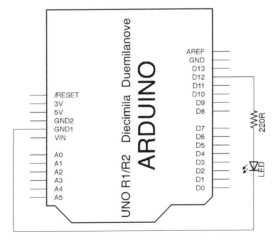

A schematic of an LED connection through a digital pin

Breadboard connections diagram

Once you have understood what we pretend to connect to the Arduino board, it's time to make the real connections using a breadboard.

Here, you have a diagram where you can clearly see the connections that should be made to implement the circuit on the breadboard. All these diagrams have been made using Fritzing, which is an excellent free application that you can find at `http://fritzing.org`.

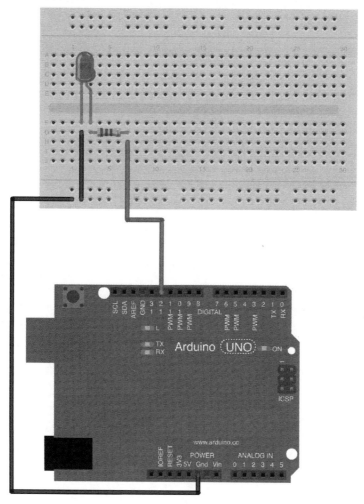

A diagram of a connection to the LED assembly in a breadboard

In this diagram, you can appreciate something that you should get accustomed to as a good practice when assembling circuits on a breadboard and that you will see in all the diagrams in this book.

As you can see, in the preceding diagram, I have connected a wire from the Arduino **GND** pin to the first hole of a rail in the breadboard, and from there, I've used another wire to connect that rail's row to the LED cathode. I haven't directly connected the LED cathode to the Arduino **GND** pin, although I could have done so.

This way, and given that all holes in every rail's row are interconnected, I have all the rail's connections tied to **GND**, which allows for a very fast and convenient way to connect the negative side of additional components to **GND**. This is done without the need for a wire from every component's negative side to the Arduino **GND** pin, which will be impossible to achieve given that there are only two **GND** pins in the Arduino board.

Later in the book, we will also need to power other electronic components with a positive voltage, and we will use the same technique shown here, that is, we will take a wire from the Arduino 5V pin to another row of a rail, thus getting a whole row with positive voltages ready to power additional components on the breadboard.

Asymmetric blinking code

Once we have assembled the circuit and connected it to our Arduino board, it is time to leave the physical part of our project and begin to work on the logical one: the software we will program our microcontroller with and that will allow us to command the physical side of our project.

For this first example, we will use a modified version of the Blink example you saw in the previous chapter, and will blink the LED in an asymmetrical pattern, that is, having different durations for the on and off cycle of the blinking:

```
void setup() {
  pinMode(12, OUTPUT);
}

void loop() {
  digitalWrite(12, HIGH);
  delay(500);
  digitalWrite(12, LOW);
  delay(100);
}
```

Although you have the complete code in the code examples accompanying this book, I'd suggest that you type it by yourself, because this is the only way to really learn how to code and try to strictly respect the C programming language syntax.

This simple code will help us to analyze and understand the common structure of an Arduino sketch.

As you can appreciate, the code is divided into two different sections called functions. To be precise, we have a function called `setup()` and a function called `loop()`, and each one has a very concrete mission in a sketch:

- `setup()`: The purpose of this function is, as its name indicates, to set up the board and its peripherals in the way the sketch needs, for example, setting pins as inputs or outputs, assigning a predefined value to a variable, or initiating a serial communication with your computer. It's only executed once — right at the beginning of the program.
- `loop()`: This function is where the real code execution occurs. It runs through every instruction within it in a sequential way from the top to the bottom and begins again once it has reached the last instruction. This function never finishes its execution, so the only way to stop running a sketch in Arduino is by powering the board off.

The logic of the code is quite simple. In the `setup()` function, we just tell Arduino that we want to use pin **12** as an output using the `pinMode(12, OUTPUT)` function call that, as you can see, takes two parameters:

- The first is the pin we want to configure
- The second is the mode we want it to be in, which is `OUTPUT` in our case

Once the setup is finished, Arduino enters the `loop()` function.

The first thing we do there is write a digital `HIGH` value in pin **12** using the `digitalWrite(12, HIGH)` function, which generates a 5V signal on the pin, making the LED turn on.

After that, we wait for 500 milliseconds with the call to the `delay(500)` function and go on by writing a digital `LOW` value on the pin with the call to the `digitalWrite(12, LOW)` function, thus generating a 0V signal that makes the LED turn off.

Finally, we wait again but only for 100 milliseconds with `delay(100)` in this case, just before repeating all the loop code again, that is, turning the LED on again, waiting for half a second, turning it off, and waiting for a tenth of a second and again forever until we power the board off. Simple, isn't it?

Here, you have a picture of the complete circuit connected to the Arduino board. Please note that for the sake of clarity in the photograph, in the final assembly I decided to save one wire by directly connecting the LED cathode to the Arduino **GND** pin and not connecting it through the rail, as presented in the connections diagram previously.

The complete circuit for the asymmetrically blinking LED

C language syntax considerations

This being our first code, I would like to consider the syntax of the C language used in it. This, once understood, will help you minimize the syntax mistakes you could make when writing C code, and that would generate compiling errors. They are as follows:

- The C language is case-sensitive; this means that it is not the same as a word written in uppercase or lowercase: `pinMode()` is correct whereas `pinmode()` isn't.

- The use of functions includes the accompanying pair of parentheses that serve to specify their parameters. Even when they don't have any parameters, the parentheses have to be included.

- Every function content or block of code requires the use of a pair of curly brackets to delimit the instructions that belong to them and separate them from the rest of the program. Missing a closing curly bracket would result in a compilation error, which is sometimes hard to detect and correct.

- As you can see, there is a semicolon at the end of every instruction. They are required by the C language to know where every instruction ends and where the next one begins. A simple carriage return isn't enough.

- Everything after `//` or between a `/* ... */` block is considered to be a comment and thus, it won't be compiled by the Arduino development environment. The use of comments to explain the code is a common and very desirable practice that you should adhere to.

Troubleshooting faults in the circuit

It's difficult to have any trouble with such a simple circuit, but who knows, and perhaps there is something wrong in your assembly and you don't get the expected results. In the case of problems, here is a list of things you should check:

- Go over your connections carefully, especially in the breadboard. The clips under the plastic housing of the breadboard tend to open wide gradually and sometimes, even when the component or wire is inserted in the hole, it may not make a correct contact.

- On the Arduino side, be sure to connect the wire going to the resistance to pin **12**, which is the one you are referencing in your code to be a digital output and the one you are using to turn the LED on and off. It wouldn't be the first case, as I've spent plenty of time looking for an error in the breadboard's connections when it was just that I had connected the wrong pin in the Arduino board.

- The LED is a polarized component. This means that its legs have different functions. The anode usually has a longer leg and the cathode has a small flat in the plastic capsule to differentiate them. Ensure that it is connected the right way, that is, the anode is connected to the resistance and the cathode to the wire going to the ground rail.
- If it still doesn't work, try with another LED, unwire everything, and connect it once again or try with another breadboard.

Dealing with multiple outputs

Once we have our first real circuit up and running, or blinking if you prefer, and once you are a little bit acquainted with the structure of a typical sketch and the C language syntax, why don't we try to make something a little more complicated?

It isn't hard to connect two more LEDs and their corresponding current limiting resistances and build a traffic light and modify our sketch to make it operate like a real one.

Here, you have the schematic of such a circuit:

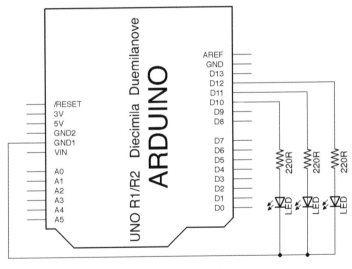

A traffic light circuit

Interacting with the Environment the Digital Way

As you can see, I have connected the resistances to pins **10**, **11**, and **12**. Pins **10** and **11** are marked as **PWM** capable, but for the purpose of this circuit, this doesn't really matter to us, because we are going to use them digitally by calling the `digitalWrite()` function later in the code.

In the following breadboard connections diagram, you can see what we talked about previously regarding the creation of a ground rail that would allow us to connect every ground in the circuit in a convenient way. You can see how I've taken different wires from the cathode of every LED to the common ground rail at the bottom.

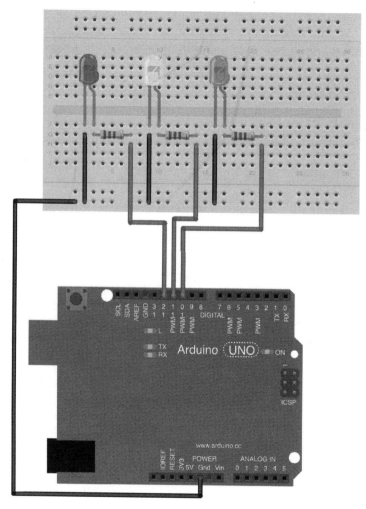

A breadboard connections diagram for the traffic light circuit

Current limit per pin

There is a very important consideration to be made before going into connecting bigger external circuitry to our Arduino boards.

The Arduino board is only capable of delivering a maximum current of approximately 40 mA per pin and always under a total current delivery of 200 mA for all of them at the same time. Above these limits, your Arduino board could be seriously damaged and get burnt.

In this circuit, with a total of three LEDs and with a power consumption of around 20 mA each, we are sure we won't get into any trouble.

If you have to deal with devices that require higher power consumption, you should consider using a relay or a transistor as an intermediation between the Arduino board and an external power source for the device. In *Chapter 4, Controlling Outputs Softly with Analog Outputs*, I'll show you how to use a transistor to control a DC motor for this very reason.

Coming back to our example, the code for this circuit won't be much harder than the blinking example used previously, but I'll use it to introduce a very important concept of any programming language: the use of variables to store values that we will use later in the code. In this case, I'll use these variables just to make the code easier to read, but for the purpose of our example, it will be enough.

I'll insert three lines of code before the `setup()` function that will allow us to declare three variables that we will use to store the pin numbers we will reference later in the rest of the program:

```
int redLED = 12;
int yellowLED = 11;
int greenLED = 10;
```

In the `setup()` function, we are going to configure the pins we want to use by making them all outputs:

```
pinMode(redLED, OUTPUT);
pinMode(yellowLED, OUTPUT);
pinMode(greenLED, OUTPUT);
```

Interacting with the Environment the Digital Way

Finally, in the `loop()` function, we are going to just repeat three times what we have done before, but we'll modify the code so that it turns a different LED on and off each time:

```
digitalWrite(redLED, HIGH);
delay(500);
digitalWrite(redLED, LOW);
delay(100);

digitalWrite(yellowLED, HIGH);
delay(500);
digitalWrite(yellowLED, LOW);
delay(100);

digitalWrite(greenLED, HIGH);
delay(500);
digitalWrite(greenLED, LOW);
delay(100);
```

Here, you have the complete code including some comments so that you can clearly read what is going on:

```
/*
 Chapter 03 - Interacting with the environment the digital way
 Multiple digital outputs to simulate a traffic light
 By Francis Perea for Packt Publishing
*/

// Global variables
int redLED = 12;
int yellowLED = 11;
int greenLED = 10;

// Configuration of the board.
```

```
// All pins are going to be used as outputs
void setup() {
  pinMode(redLED, OUTPUT);
  pinMode(yellowLED, OUTPUT);
  pinMode(greenLED, OUTPUT);
}

// Sketch execution loop
// We repeat the single blink for every LED
void loop() {
  // blink the red LED
  digitalWrite(redLED, HIGH);
  delay(500);
  digitalWrite(redLED, LOW);
  delay(100);

  // blink the yellow LED
  digitalWrite(yellowLED, HIGH);
  delay(500);
  digitalWrite(yellowLED, LOW);
  delay(100);

  // blink the green LED
  digitalWrite(greenLED, HIGH);
  delay(500);
  digitalWrite(greenLED, LOW);
  delay(100);
  // do it all again
}
```

Interacting with the Environment the Digital Way

Finally, here, you have a photograph of a real assembly, where, once again, I made use of a little trick to save some wires and electronic components—resistances, in this case. Given that we are not going to turn on more than one LED at a time, I've just used a single resistance connected to the cathode of every LED. All LEDs will use it, but they'll use just one at a time, so we can save two resistances. On the other hand, having just a single connection to ground on the opposite leg of the single resistance, I can also save some wires to connect it to the Arduino ground—once again, just for the clarity of the picture.

A complete assembly of the traffic light example

Now that you are beginning to master digital outputs, will you be able to make the LED in the middle blink a couple of times before turning it off and prior to turning the lower one on like we usually see in our streets' real traffic lights?

Summary

This has been our first real work project, and I've shown you quite a good bunch of concepts and procedures.

We have considered the breadboard as our sandbox, where we will try to connect every circuit we are going to build throughout the book and in your daily work with Arduino.

We have connected our first component to the Arduino board through a digital output and have met the power connections of our circuit.

Finally, we have had our first real contact with an Arduino program written in the C programming language. We have recognized the essential parts of any Arduino sketch and learned the basic syntax of the C language and its most important particularities.

In the next chapter, we will learn about analog outputs, which will introduce us to a very powerful transistor-based circuit that will allow us to deal with no less than a DC motor. On the programming side, we will meet another very interesting structure to step along a range of values. This is getting interesting, isn't it?

4
Controlling Outputs Softly with Analog Outputs

Interacting with the environment in a digital way is very practical, and you'll use it in a different number of situations. In fact, we are very accustomed to these kinds of interactions and they are nothing really new.

In this chapter, we will see a very different kind of interaction that not only supports an on and an off state but also a number of different states between them. We will deal with analog outputs.

We will begin by fading an LED to later see a new circuit that controls the speed of a motor through an analog output.

Dealing with analog signals

As an introduction to the connection and programming of analog outputs, we will use the simplest circuit we have already seen, that is, we will connect an LED to our Arduino board but to deal with it analogically this time.

Before going into the details, I would like to introduce a couple of concepts that will help you when working with these kinds of signals.

The first thing you should know is that Arduino isn't really able to generate an infinite continuous analog signal, but instead, it uses a little trick to simulate it. Digital devices such as microcontrollers usually incorporate a peripheral called **Digital to Analog Converter (DAC)**, which they use specially to perform this trick.

When working with a microcontroller, we pass a digital value to the DAC, and it converts it to an analog value but in a predefined range of possible values. The DAC is unable to generate an infinite set of output values; it has a finite input range of digital values and can generate a finite output range of analog voltages.

The number of steps the DAC can generate is called the resolution of the DAC. The Arduino DAC has an 8-bit resolution, so it accepts input values ranging from 0 to 255, which will be converted to analog values between 0V and 5V. If you divide the voltage range between the total number of steps the Arduino DAC accepts, you will conclude that every step of the input range increments the output voltage by almost 0.02V.

To perform this conversion, the DAC uses a method called **Pulse Width Modulation** (**PWM**) that consists of turning the output at very high frequencies on and off, resulting in a medium voltage that is the proportion of the on time with respect to the off time.

If you want to learn more about DACs and PWM, you can visit the corresponding pages on Wikipedia at http://en.wikipedia.org/wiki/Digital-to-analog_converter and http://en.wikipedia.org/wiki/Pulse-width_modulation.

The analog output circuit

Once we know what is going to happen when we deal with analog outputs, I'll show you the circuit we are going to work with.

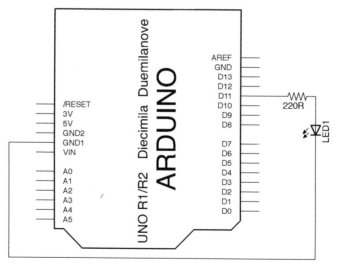

The circuit to connect an LED to a PWM pin

As you can see in the previous schematic, it's very similar to the circuit we used in *Chapter 3, Interacting with the Environment the Digital Way*, to deal digitally with an LED. However, in this case, we have to connect it to a **PWM** pin of the Arduino, pin **11** for this example, so that later in the program we can use the corresponding code to deal with it analogically.

The rest of the circuit is just the same as the blinking LED one, and we will use the same 220 Ohms resistance to limit the current that will flow through the LED to no more than 20 mA.

Connections diagram

If you still see connections better in the connections diagram for the breadboard, here you have the complete diagram for this example:

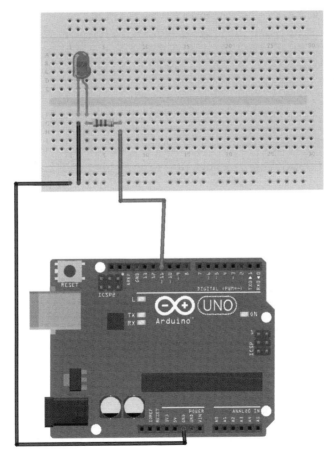

The connections diagram for the fading LED

Analog control through code

The most valuable part of this example is not its physical part but its logical one, and I'll show you the associated functions to deal with analog outputs and a very important control structure in any programming language called a `for` loop in the code we will be using for this circuit.

The analogWrite() function

Dealing with analog outputs from the programming side is just a matter of using the `analogWrite()` function, which, as its name implies, writes an analog value through a pin.

It takes two parameters, similar to the `digitalWrite()` function:

- The first one is the pin on which we want to operate (remember that it has to be one marked as **PWM**) on
- The second parameter is the value we want to output, and that, as I told you previously, can take any value in the range of `0` to `255`, `0` meaning 0V and `255` meaning 5V

The for loop

The other important concept we will come across in this example is a `for` loop, which is a very important control structure present in almost every programing language that allows you to repeat a block of code a specific number of times.

For its operation, a `for` loop needs what is usually called a control variable, that is, a variable that is going to take different values in a range for every iteration of the loop until it reaches a final value.

The general syntax of a `for` loop is as follows:

```
for (initialization; condition; increment) {
   instructions;
}
```

The three parts inside the parentheses have the following mission:

- **Initialization**: An initial value has to be given to the control variable. The loop will begin to iterate, with the control variable taking this value the first time.
- **Condition**: The loop will go on iterating as long as this condition remains true, stopping the iterations and going on with the next instruction as soon as the condition evaluates to false. The condition is usually based on the control variable remaining under or above an extreme value.
- **Increment**: This is a sentence where we change the control variable in order to continue adopting different values in the range through its way to the final value. Normally, we simply increment the control variable, but in some cases, as we will see in our example code, we could also decrement it.

An example will greatly clarify the concept. We could write the following block of code:

```
for(time = 100 ; time <= 1000; time=time+50){
  digitalWrite(led, HIGH);
  delay(time);
  digitalWrite(led, LOW);
  delay(time);
}
```

In this example code, we will begin by making the time variable take a value of 100 and make the loop iterate, executing all instructions inside as long as the time variable has a value less than or equal to 1000, incrementing it by 50 in every iteration.

This way, we would make an LED blink with lower frequencies for every iteration, given that we have used the time variable as the parameter of the delay() calls.

It would result in delay(100) as the first iteration, delay(150) as the second one, delay(200) as the third one, and so on until the time variable gets to a value of 1000, in which case, the condition will evaluate to false and the for loop won't continue to iterate. The program execution will follow through the next line of code just after the closing curly bracket of the for loop.

For more information and some other examples of the use of a for loop, you can go and visit the for loop page on the Arduino site's **Reference** section at http://arduino.cc/en/Reference/For.

Complete the fading LED code

Regarding the rest of the code, I'm going to use three variables; two of them, which I've called `led` and `increment`, won't really change throughout the execution of the program, but they will greatly increase the readability of the code.

The `led` variable is going to be used to store the pin number I'll use to control the LED, pin number **11**, in this example. Remember that you can use whichever pin you want from those marked as **PWM** in the Arduino board and that you will have to set this variable as per your choice, accordingly.

The `increment` variable will hold the step size we will use in the `for` loops that we will use later. Using a variable just to use the `increment` value makes it very convenient to make future modifications by just changing its value at the beginning of the code without the hassle of having to search throughout the code for every occurrence of the pretended value.

The other variable, which is `intensity`, will be the key of this sketch, as it is going to be used as the control variable of two `for` loops that we are going to use to go all along the range of possible values for the `analogWrite()` calls we will use to gradually change the brightness of the LED.

Well, let's see this theory in action. Here is the complete code for the LED fading sketch:

```
/*
 Chapter 04 - Controlling outputs softly with analog outputs
 Single analog output to fade a LED
 By Francis Perea for Packt Publishing
*/

// Global variables we will use
int led = 11;
int intensity = 0;
```

```
    int increment = 5;

    // Configuration of the board
    void setup()  {
      // Set the pin we are going to use as an output
      pinMode(led, OUTPUT);
    }

    // Main loop
    void loop()  {
      // fade from minimum to maximum
      for(intensity=0; intensity<=255; intensity=intensity+increment){
        analogWrite(led, intensity);
        delay(30);
      }

      // fade from maximum to minimum
      for(intensity=255; intensity>0; intensity=intensity-increment){
        analogWrite(led, intensity);
        delay(30);
      }
    }
```

As I have explained previously, we begin by declaring and assigning initial values to the three global variables we are going to use all along the program.

In the `setup()` function, we simply set the pin we are going to use as an output.

In the `loop()` function, we have just two `for` loops: one taking values from 0 to 255 in increments of 5 and the other taking values from 255 to 0 using -5 increments.

In every iteration of each `for` loop, we simply set the changing value of the control variable, which is `intensity` in our code, as the analog value we want to output to the LED and wait for 30 milliseconds to allow our eyes to see the change.

The first loop makes the intensity of the LED increase from 0V to 5V and the second one makes the intensity decrease from 5V to 0V, which is just the opposite.

I'll suggest that you play with this code, and as an exercise, you could make both parts of the fading run at different frequencies simply by setting different values for the `increment` variable just before entering the `for` loops, making it `10` when going up and `50` when going down. The resulting code could be as follows:

```
void loop()  {
  // fade from minimum to maximum in increments of 10
  increment = 10;
  for(intensity=0; intensity<=255; intensity=intensity+increment){
    analogWrite(led, intensity);
    delay(30);
  }

  // fade from maximum to minimum in decrements of 50
  increment = 50;
  for(intensity=255; intensity>0; intensity=intensity-increment){
    analogWrite(led, intensity);
    delay(30);
  }
}
```

Motor control with a transistor

Analog outputs can be very useful sometimes and not just to change the brightness of an LED. There are plenty of devices that operate on an analog signal; motors, for example, where you can change its speed by varying the voltage you apply to them.

However, motors can be sometimes tricky to operate, mainly due to the fact that they are big current consumers. A typical toy DC motor can easily consume more than 200 mA when running without a load and up to 1 Amp when stalled.

We mentioned in the previous chapter that an Arduino pin can't give more than 40 mA, or it could burn. So, how can we deal with a motor using an Arduino? Well, usually when dealing with high-current devices, we use a **driver** circuit that allows the device to be powered from an external power source and use the Arduino pin just as a regulator.

This way, we can avoid the need to power the device directly from the Arduino pin, like we have done till now when dealing with LEDs that operate with under 30 mA and can be directly fed from Arduino pins.

Motor driver

This kind of circuit usually uses a transistor as an intermediate device that receives the Arduino signal and provides a proportional current coming from an external power source to the device.

In the following schematic, you can see a typical assembly when operating a DC motor from Arduino:

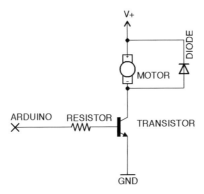

A motor driver circuit

The key element of this circuit is the transistor component, the one with the arrow pointing out that has three legs:

- **Base**: This is the leg that we connect to the Arduino board through a resistor and that acts as the control element
- **Collector**: This, simply said, is the leg where current comes into the transistor
- **Emitter**: This is the leg through which current flows out of the transistor

Regarding its operation, I like to compare a transistor with a water tap, where we turn the handle and, more or less, water comes out of it. In the transistor case, we set a variable voltage on the base, which allows more or less current to flow from the collector to the emitter and consequently, through all components we connect in between.

Keeping this in mind and taking a closer look at the previous schematic, we can realize that by connecting the base to an Arduino analog output, we can make a varying current flow through the motor, thus making its speed vary.

When dealing with transistors, it is important that you know which of the three legs is the base, collector, and emitter of your transistor. You should refer to its datasheet, which you can easily find on sites such as `http://octopart.com`.

There are two more components on the circuit that deserve a short explanation:

- **The resistor**: This allows the connection between the Arduino pin and the transistor base. Without going into too much detail, we will use a 1K resistance here.
- **The diode**: Without going into too much detail, you should know that a motor acts much like a big electromagnet and thus, when turned off, generates what is called a back **Electromagnetic Force** (**EMF**), which can even be 100 volts and with an opposite direction from the main voltage applied to the motor, which could damage your circuit. By placing a diode in the opposite direction of the main voltage, you help suppress this dangerous back EMF. Any protection diode will do, such as the 1N4007, but if you don't have one to hand, you could even use an LED as long as you connect it correctly.

Power source considerations

There is a last question to consider before going hands on with our motor control circuit, and that is the powering of the motor.

For the purposes of this example, I'll suppose that you are going to connect a small DC motor that consumes no more than 250 mA when operating without a load. This way, we could take current directly from the 5V pin of the Arduino board.

It's important to keep this in mind, because the power regulator present in your Arduino board takes current in the last instance from your USB port, which can't typically support more than 500 mA.

You should take some precautions before connecting a bigger motor to your Arduino board by measuring its typical power consumption. You can do this by placing an ammeter in the serial with a battery or another external power source and take note of the current it needs to run.

If the motor doesn't need more than the maximum 500 mA that your USB port and Arduino voltage regulator can give, then you can safely power it from the 5V Arduino pin. In other cases, you will have to provide another power source for the motor and connect both grounds: the one for your external power source and the one for your Arduino board.

The complete circuit

Well, that is enough theory for the moment; let's go into the real nuts and bolts of our circuit. Here, you have the complete schematic of the motor speed control circuit we are going to assemble for this example:

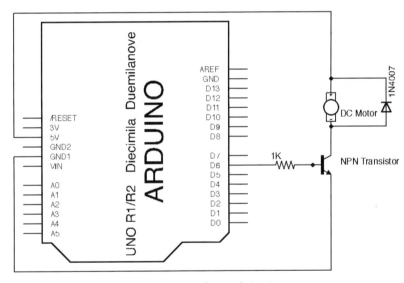

A motor speed control circuit

Connections diagram

For the breadboard connections, we can save some wires by placing the diode just in contact with the transistor collector.

Notice that for the transistor used in this diagram, the legs are looking at it with the plain side in front from the left to the right, collector, base, and emitter, or, as you will usually find, CBE.

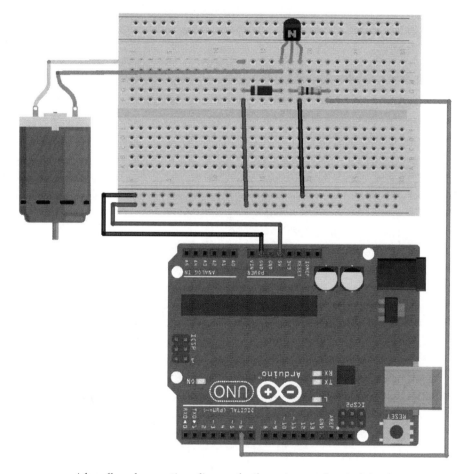

A breadboard connections diagram for the motor speed control circuit

Motor varying speed code

The code for this circuit is very similar to the LED fading one, but instead of using the for loop to go all over the range of values of the analog output, we will just use it to generate a three-iterations loop that will give us just three different speeds for the motor, which will make its changing state more evident.

Further, I couldn't make my motor spin with a voltage lower than 1V, so I will configure the loop to take the values 150, 200, and 250, which made my transistor deliver voltages of 1.5V, 2.5V, and 3.5V.

Here, you have the complete code I used for this example:

```
/*
  Chapter 04 - Controlling outputs softly with analog outputs
  Motor speed control
  By Francis Perea for Packt Publishing
*/

// Global variables we will use
int base = 6;
int speed = 0;

// Configuration of the board
void setup() {
  // Set the pin that we will connect to the transistor base as an output
  pinMode(base,OUTPUT);
}

// Main loop
void loop() {
  // Increment the speed of the motor in three steps, each for 3 seconds
  for (speed=150; speed<=250; speed=speed+50){
   analogWrite(base, speed);
   delay(3000);
  }
  // Stop the motor for 1 second and begin again
  analogWrite(base, 0);
  delay(1000);
}
```

Notice that every change in speed lasts three seconds and that once the three steps' loop ends, we stop the motor for one second before beginning all over again.

The assembled circuit

I'd like to show you a picture of the complete assembly that could perhaps clarify even more what we are doing:

A real assembly for the motor speed control circuit

As you can see, I've used a little piece of tape around the motor axle in order to make it clear when it changes its speed.

Bigger power motors

As I told you previously, when dealing with motors, this can consume more current than the maximum 500 mA your USB port can give. You will have to provide another power source just for the motor, such as an external battery or DC transformer.

I have used an excellent and totally recommendable free software called EAGLE (http://www.cadsoftusa.com) to prepare a schematic for this kind of circuit that can perhaps help you understand what we pretend.

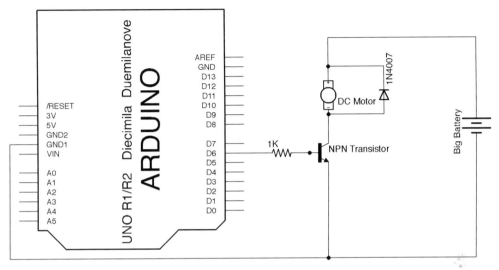

A circuit to power the motor from an external power source

Here, you can see how the motor gets its current from the big battery and the only component between both circuits is the transistor, which is connected through its base to the Arduino board, its collector to the external power source, and the emitter to both grounds: the battery one and the Arduino one.

This way, you can not only provide a bigger current but also a different voltage to the motor that could, for example, operate at 12V as opposed to the 5V of the Arduino board.

One final consideration with respect to this circuit is that you should look for an appropriate transistor, because every transistor has a maximum current that it can support, so be careful if you don't want to see a little fireworks show on your desktop.

Summary

In this chapter, we met analog outputs and saw two different circuits to deal with two kinds of components that can operate on a variable voltage, an LED, and a motor.

On the hardware side, these circuits have helped us to learn about analog signals, DACs, and some of its characteristics. We also learned what a driver circuit is and how to use a transistor to create one, helping us know something more about this component. We even saw a very practical way to deal with high currents from Arduino without providing them from the Arduino board voltage regulator.

From the logical point of view, we met the `for` loop, a very handy control structure that has helped us go over ranges of values and loop iterations.

By now, there's been enough about outputs, and in the next chapter, we are going to begin to work with inputs—digital inputs, to be precise, which will serve us as the base of all kinds of sensor connections. This will open us to a totally new way of looking at microcontrollers. Are you ready?

5
Sensing the Real World through Digital Inputs

Dealing with outputs is really just the half of the pie. Inputs are the other half, and if I may, I'd say they are the most interesting part of any project you will work on. In this chapter, I'll introduce you to the use of digital inputs as a way to make your project sense their environment through the assembly and programming of two different examples that I'm sure you will like.

Sensing by using inputs

Inputs, whether they are digital or analog, are the way through which Arduino can sense what is happening around it. In some cases, they are used as an interface with the user, such as when we connect buttons and switches. In other cases, we use inputs to measure a physical variable that will make our project react in some way or other.

Nowadays, we can find almost a different sensor for every physical variable we want to measure and an endless number of devices that allow humans to interact with electronic devices. So, the first thing we will have to consider is the type of sensor we will need for our project and, most importantly, from the point of view of connection and programming, the type of signal it generates, digital or analog. This is because it will affect the way in which we have to connect it to our Arduino board, by way of a digital or analog input, and program our sketch accordingly.

Digital sensors usually give us a 0V or 5V voltage or any way to obtain it, which can be easily traduced into HIGH or LOW values in the Arduino code. In other cases, they provide a different value that we will have, by means of additional circuitry, to adapt to the Arduino digital input allowed voltage, that is, 0V to 5V.

Regarding analog sensors, their main characteristic is that they provide a continuous value between a range, and we will also have to adapt them to our Arduino allowed range, but we will talk deeply about analog inputs and sensors in *Chapter 6, Analog Inputs to Feel between All and Nothing*.

At the moment, let's see the simplest digital input circuit, a switch, and how to program it.

Connecting a button as a digital input

The simplest circuit that we can prepare to be used as a digital input is that consisting of a switch connected to a digital input of the Arduino board.

A typical momentary push button usually has two states, open or closed, and acts like a switch that, when not pressed, keeps the circuit open, preventing the connection between the two parts of the circuit it connects, and when pressed, makes these two parts connected.

Keeping in mind that an Arduino digital input can sense values of 0V and 5V, we have to prepare our circuitry in such a way that it provides a 0V signal to the Arduino digital input when we want to consider it LOW and a 5V signal when we want to consider it HIGH.

To be precise, the Arduino microcontroller can sense anything between 3V and 5V as a HIGH value and something between 1.5V and 0V as a LOW value, the range between 1.5V and 3V undetermined and, thus, not valid.

For the purposes of our first digital input example, we will use a very simple circuit that you can see in the next schematic:

Chapter 5

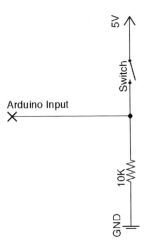

A basic switch connection to Arduino

Although it is a simple circuit, it may deserve some explanation. It basically exposes two different states: the button or switch pressed or released:

- **Released**: When the button is not pressed, the two parts of the circuit remain unconnected, thus not flowing any current through that branch of the circuit. In this case, the Arduino board only senses a GND or 0V value through the digital input by which we connect this circuit to it due to the fact that it is physically connected to the ground through the resistor.

- **Pressed**: When we press the button or close the switch, what we are internally doing is connecting both parts of the circuit, that is, the 5V upper side to the GND lower side, allowing current to flow from one to the other through the 10K resistor. This is precisely why we use the 10K resistor: to provide some load to this branch of the circuit, preventing the short circuiting of the power source. From the point of view of the Arduino board, it senses a 5V signal in the pin we use to connect this circuit to the board.

The momentary push button

In the next screenshot, you can see a bunch of different momentary push buttons of two kinds:

- **Printed circuit board (PCB) soldering**: These used to be smaller, really tiny in some cases, and they come with little legs that allow them to be soldered through holes in a PCB. In the screenshot, they are the three on the left-hand side.

- **Panel mounting**: These are the three on the right-hand side in the picture. Usually, they come in a much bigger size than the PCB kind and come with bigger connections so that wires can be soldered to them. They usually come with some kind of nut so that they can be fixed to a panel.

Different momentary push buttons

Physical differences apart and looking at them from the connection point of view, we have to notice an important difference between both groups.

If you take a closer look at the PCB group or, even better, if you perform a simple search on the Internet, you will notice that these kinds of momentary buttons usually come with a total of four legs instead of simply two. Usually, the reason is to give a stronger union to the PCB group by allowing up to four soldering points, but also because they have their connections duplicated to provide a more reliable connection between both points of the switch.

This means that usually, momentary push buttons with four legs have them internally connected two by two, which looks like what is shown in the next diagram:

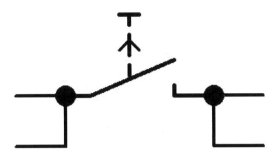

An internal wiring connection of a typical momentary push button

This simple feature has to be taken into serious consideration when connecting one of these buttons through a breadboard, because if we connect them in the wrong way, we could easily short-circuit the Arduino board, with subsequent bad consequences.

The first thing you should do when using one of these momentary push buttons is to identify every leg so that you know which of them are internally connected. They usually come with some kind of indication or diagram, but if they don't, you could always use a voltmeter or continuity tester to find out.

Once you have determined which legs are connected, you should always connect the momentary push button just above the centerline of the breadboard, which divides the two groups of holes in such a way that you leave two of the replicated connections out of use and ensure that the other two corresponding to the two contact points of the switch are always connected in different columns of the breadboard connections block. This way, they prevent a short circuit when you connect them to the positive rail (5V) and ground.

Also remember that this kind of circuit needs a current limiting resistor in a part of the branch that gets closed between V+ and the ground to prevent a short circuit again. A value of 10K is more than sufficient.

Complete circuit schematic

Here, you have the complete circuit schematic for this example. As you can see, the connection point between the switch and the resistor is connected to the Arduino pin number **7** in this example, and we will use pin number **9** as an output to connect the LED that we will use to blink or fade.

Given that we will deal with the LED as a digital or analog device up to the pressing of the button, we have to connect it to the Arduino board through any of the PWM pins that will allow us to use it with the `digitalWrite()` or `analogWrite()` functions.

The only consideration to be made about this circuit is regarding the right connection of the momentary button we will use, and we have talked about this in the previous paragraph.

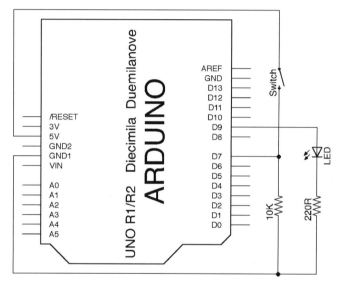

A blink or fade circuit schematic diagram

Breadboard connections diagram

All that said, I'm sure you will see it clearly in the breadboard connections diagram, where you can notice how the two upper legs of the momentary push button are left on the upper-side connections block of the breadboard without connecting anything to them and separating them from the other two legs, thanks to the central space of the breadboard.

You can also see how once we push the button, the current flows from the positive rail to the ground through the switch and the current limiting resistor, and it is in just this point, after the switch and before the resistor, where we take a wire to the Arduino pin that we will use as a digital input.

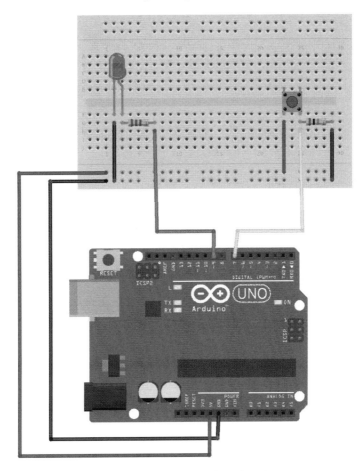

A breadboard connections diagram of the digital input circuit

Writing code to react to a press

Before we go into the details of the programming of such a circuit, we'll have to decide what behavior we will unchain in our project once the event of a press is detected. For the sake of simplicity, we will make an LED blink or fade depending on the press of the button.

This way, we can test all that we have already learned in a simple example and this will also lead us to meet another very important control structure of any programming language, called a conditional bifurcation, which some say is the basic structure that makes any program a logical structure and not a simple sequence of instructions.

Once the circuit is assembled, let's take a look at the code we are going to write to detect and react to the button press as a digital input. Here, you have the complete code for this example:

```
/*
  Chapter 05 - Sensing the real world through digital inputs
  Sensing a switch
  By Francis Perea for Packt Publishing
*/

// Global variables we will use
int led = 9;
int button = 7;
int pressed = 0;
int intensity = 0;
int increment = 10;

// Configuration of the board: one output and one input
void setup() {
  pinMode(led, OUTPUT);
  pinMode(button, INPUT);
}

// Sketch execution loop
void loop(){
  // We read the button pin
  pressed = digitalRead(button);
  // if it is not pressed lets blink digitally
  if (pressed == LOW) {
    digitalWrite(led, LOW);
    delay(50);
```

```
      digitalWrite(led, HIGH);
      delay(50);
   }
   // otherwise lets fade analogly
   else {
      for(intensity=0; intensity<=255;
intensity=intensity+increment){
         analogWrite(led, intensity);
         delay(30);
      }
      for(intensity=255; intensity>0; intensity=intensity-
increment){
         analogWrite(led, intensity);
         delay(30);
      }
   }
}
```

Configuring and reading a digital input

From the point of view of programming, there are basically two main new concepts in this example regarding the use of a digital input:

- `pinMode(button, INPUT)`: In the `setup()` function, we have to set the pin we are going to sense as an input so that Arduino can read from it. The function is just the same `pinMode()` function we have been using so far, but instead of using the `OUTPUT` constant, this time, we use `INPUT` to set the pin accordingly.

- `digitalRead(button)`: When we deal with an input whether it is a digital or analog one, we are going to receive the read value instead of setting a value like how we have been using outputs. This means that we only pass a parameter indicating the pin to be read to the reading function, `analogRead()` in this case, but it also means that this function, as opposed to writing functions, is going to return us a value that we have, in some cases, to store in a variable for later use.

Being realistic, in this example, it isn't really necessary to store the read value in a variable because we are not going to use it anymore in the rest of the code, and we could simply have used the `digitalRead()` function inside the `if` parentheses like `if(digitalRead(button) == LOW)`, but for the sake of clarity in this first example, I preferred to use a variable this time.

Taking decisions with conditional bifurcations

Beside the differences between the use of inputs and outputs, there is an even more important concept in this example, and it is the use of the `if` control structure to decide what has to de done depending on the state of a previous event.

The `if` control structure is the basis of any programming language, giving them the power to decide and act on consequences.

The simplest syntax of the `if` sentence is as follows:

```
if (condition) {
   Block of instructions to be executed
}
```

Being a condition, any logic expression is one that evaluates as true or false and that can use logical operators such as `==` (is equal to), `>` (is bigger than), `<` (is less than), or `!=` (is different).

In a more complex format, the `if` control structure can even include a block of instructions to be executed in case the condition evaluates as false, in which case, its syntax is as follows:

```
if (condition) {
   instructions to be executed in case the condition evaluates as true
}
else {
   instructions to be executed in case the condition evaluates as false
}
```

As you can see in the code for our example, this format is the one we have used to get two different reactions on the pressing of the button:

- If the `pressed == LOW` condition is true, being pressed the value read from the digital input and meaning no press on the button, we execute a simple blink
- In the case of the condition being false, which means that there has been a press on the button, we execute the block of instructions contained in the `else` branch of the code

For a deeper reference with respect to the `if` control structure in the Arduino programming language, you can visit the Arduino website's **Reference** section, particularly the pages related to `if` and `if ... else` at http://arduino.cc/en/Reference/If and http://arduino.cc/en/Reference/Else.

Timing and debouncing

If you play a little with the previous circuit once programmed, you can easily conclude that the reaction to the button press is not immediate, and it is due to the fact that in the code, the button press is only read once the reaction, blinking or fading, has concluded and both of them take their time, even if short.

In *Chapter 9*, *Dealing with Interrupts*, we will see another way to react immediately to these kinds of events through the use of a very powerful procedure called interrupts.

There is another final consideration with respect to this example that is called debouncing, and that is due to the physical way in which electrical contacts are made when pressing a button.

Although it's hard to imagine, when you press a button, the state oscillates between on and off before the contact settles down, which can lead to incorrect readings of a sensor in the code.

You can learn more about switches and debouncing them on the Wikipedia page for switches at `http://en.wikipedia.org/wiki/Switch`.

The Arduino site also has a very interesting page exclusively dedicated to buttons debouncing at `http://arduino.cc/en/pmwiki.php?n=Tutorial/Debounce`, which deserves a reading.

Other types of digital sensors

Switches are perhaps the most used devices as digital input sensors but obviously not the only ones. Any device that can open or close a circuit branch can be easily configured and connected to Arduino to act like a digital input sensor, from a reed relay (`http://en.wikipedia.org/wiki/Reed_relay`) to a PIR motion detector (`http://en.wikipedia.org/wiki/Passive_infrared_sensor`).

One of those devices that are very cheap and easy to acquire are optocouplers or, as they are sometimes referred to, optical switches.

An optocoupler is a very simple device that has two parts:

- **An infrared light emitter**: This is a simple infrared LED in a housing that directs its light emission in a precise direction
- **A phototransistor**: This is a special kind of transistor whose base is activated by the reception of light, infrared light in this case, and also in a special case that ensures that the phototransistor gets excited only with light coming from a concrete direction—that of the infrared LED on the other part of the housing

In the following couple of images, you can see the schematic of the internal structure of a typical optocoupler and a real picture of the one I used for my own assembly of the next project:

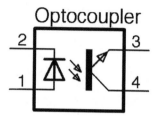 

The internal structure of an optocoupler and a real one

If you take a closer look at the picture, you may even notice the indications on the plastic housing of the emitter part with its diode sign to reference the anode and cathode and the collector and emitter markings on the phototransistor part.

If we connect the phototransistor part to our Arduino board, we could easily detect whether it is being excited or not, thus indicating whether there is something placed just in the middle of the plastic housing or not.

This was the same principle that used the first mechanical computer mice to detect and count the X and Y movement, and for that, they used a very recognizable small slotted wheel that allowed for its internal microcontroller to account for every step the mouse took in every axis.

> If you are too young, perhaps you have never seen one of this kind of mice, but you could take a look at the Wikipedia entry for mouse under the **Mechanical mice** section at http://en.wikipedia.org/wiki/Mouse_(computing).

In the next example, we will use an optocoupler as a light barrier that when interrupted, triggers a digital input.

Using an optocoupler as a coin detector

If we take advantage of the fact that there is a small space between the emitter and receptor part of a typical optocoupler that allows a coin to pass through, we could easily use one of them to create a simple coin detector that could be used for a bigger project—who knows, perhaps a candy vending machine or your next arcade game cabinet.

From the point of view of the circuit, we will have to provide little additional circuitry—only a pair of resistors: one to limit the current through the infrared LED and another to act as a load for the phototransistor between its emitter and collector.

Regarding the reaction we will get once the coin has been detected, I'll simply turn on an LED for the sake of simplicity in this example project and to allow you to focus on the digital input side of the circuit and corresponding code.

The schematic of the coin detector

In the following diagram, you can see the complete schematic for the circuit we will use in this example:

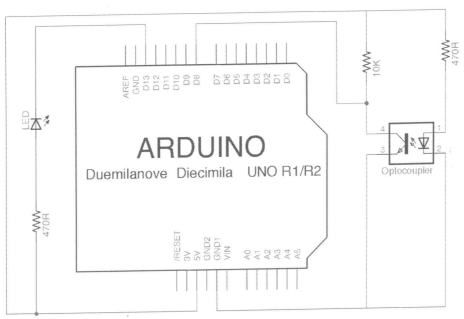

The complete circuit for an optical coin detector

As you can see, from the optocoupler side, I have only added the two resistors I mentioned previously: a 10K to act as a load for the phototransistor and a 470 Ohms one to limit the current through the infrared LED.

There is just an additional consideration. If you take a closer look, you could notice that in this case, I have connected the LED that is going to be used as the output side of our project in a different way than in other examples in the book.

In this case, I haven't connected the LED from the Arduino pin to ground but just the opposite way, that is, from **5V** to the Arduino pin, just to show you that you can also operate the LED in a negative way. This means that setting the Arduino pin to HIGH won't allow any current to flow through the LED, and setting the pin to LOW will allow approximately 10 mA to flow through the LED and into the Arduino board.

I've done it this way just to show you that it is not always necessary to provide current to output devices, but they can also can be powered from the V+ voltage source and use the Arduino pin as a virtual ground as long as the total current doesn't exceed the 20 mA limit of the total current capacity of any Arduino pin.

The breadboard connections diagram

Regarding the breadboard connections, they are a little more complex this time due to the fact that the optocoupler housing of the Fritzing part I used was a little too big for the breadboard, so I've had to use some extra wires.

That apart, please keep in mind that the exact optocoupler you may find might not have the same pin distribution as the one used in this diagram, and you should refer to your specific component datasheet to adapt the wiring connections to your optocoupler or you could easily damage it.

Chapter 5

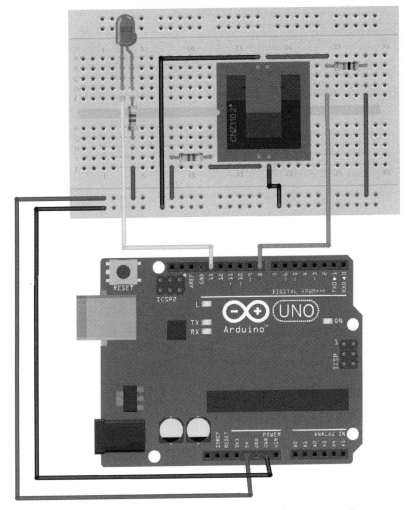

A breadboard connections diagram for the optical switch example

In the optocoupler used in the diagram, the upper connections correspond to the phototransistor and the lower ones to the infrared LED emitter.

The complete example code

There is nothing really new in the code that we will use for this example. We will, just like in the previous one, simply read the digital input corresponding to the phototransistor and only if it is a HIGH value, which means it is not receiving any light and thus not conducting, will we make the output LED blink.

Once again, just for the sake of simplicity, and given that now we know that we don't really need to store the value read with `digitalRead()` and we can simply call the functions inside the `if` condition, I have opted this time to save up this variable and simplify even a little more of the code:

```
/*
 Chapter 05 - Sensing the real world through digital inputs
 Optical coin detector
 By Francis Perea for Packt Publishing
*/

// Global variables we will use
int led = 13;
int phototransistor = 8;

// Configuration of the board: one output and one input
void setup() {
  pinMode(led, OUTPUT);
  pinMode(phototransistor, INPUT);
}

// Sketch execution loop
void loop(){
  // We read the optocoupler pin
  // and if the phototransistor doesn't receive light keep blinking
  if(digitalRead(phototransistor)==HIGH){
    //simply blink
    digitalWrite(led, LOW);
    delay(50);
    digitalWrite(led, HIGH);
    delay(50);
  }
}
```

A real working project

Finally, here you have a real picture of the complete circuit assembled with the optocoupler I had and a coin being detected:

A complete assembly detecting a coin

Summary

In this chapter, we had an introduction to the use of digital inputs by working with two different examples, each one showing different devices that can be used as digital sensors.

We saw what a momentary push button is and its particularities and also learned the right way to connect one to Arduino to be used as a digital input.

We also saw that there are lots of other devices that can be used as digital sensors and, in a second example, we used an optocoupler as an optical switch to build a coin detector.

From the point of view of programming, we met the `if` control structure in its two versions, simple and with a negative `else` branch, and learned that it can be the basis of any logical decision in our code.

In the next chapter, we are going to dive into the amazing world of analog sensors and meet two of them that will allow us to measure physical variables, which will enrich the field of application of our projects.

We will, of course, also learn how to program them and learn how to constrain their values to valid ranges in an easy way and multiply the possible execution branches of our code.

So, roll up your sleeves because this is going to get interesting.

6
Analog Inputs to Feel Between All and Nothing

We have come a long way up to this point, learning to deal with digital and analog outputs and also digital inputs. It is now the moment to show you how to manage analog inputs, as they are what enrich the microcontroller field of application the most.

In this chapter, I have prepared two projects to help you understand how to connect, configure, and program analog inputs and what kind of things we can do with them. I'm sure you will be totally amazed once you get the point.

Sensing analog values

In *Chapter 4, Controlling Outputs Softly with Analog Outputs*, we talked about digital to analog conversion and the use Arduino makes of its internal DAC when generating analog outputs. In this chapter, we need to know about the DAC cousin, the **Analog to Digital Converter** (**ADC**).

An ADC is a device with just the opposite mission of a DAC, that is, a device that takes a signal that can theoretically have an infinite number of states and convert it to just a few concrete values, or in Arduino jargon, takes an analog signal and converts it to a digital value.

Just like the DAC, the Arduino board comes with a six-channel ADC with a 10-bit resolution for each channel. This means that we can deal with up to six different analog signals ranging from 0V to 5V that will be converted to values between 0 and 1024.

If you remember from *Chapter 4, Controlling Outputs Softly with Analog Outputs*, the DAC had just an 8-bit resolution, allowing for values between 0 and 255, so we will have to keep this little difference in mind when working on projects that sense and act in an analog way, and we will have to make some kind of correspondence between the 1024 possible input values and the 255 maximum output values: a usual operation commonly called mapping.

The Arduino map function

Mapping a value from one range to another is a very simple thing. It's just a matter of finding which value will be at the same point of the output range as compared to the input range. Let's see it with a simple example.

If we have a possible input value of 0 to 100 and an allowed output range of 0 to 500 and we take a sample value of 75, we can say that it is at 75 percent of its input range, can't we? Well, which value will be at 75 percent of the output range, then? We will usually just make a simple correspondence:

Output Value = (Input Value x Maximum Output Value) / Maximum Input Value

Or, we will use the values in our example:

Output Value = (75 x 500) / 100 = 375

Here, 375 is just 75 percent of 500. Simple, isn't it?

It is so simple but also so common in the Arduino environment that we have a specific function to make this kind of mapping in the Arduino language, which is appropriately called the `map()` function.

As you can see by following the previous example, the `map()` function takes no less than five parameters: `map(value, fromLow, fromHigh, toLow, toHigh)` and returns us another value. The `map()` function parameters' meanings are as follows:

- `value`: This is the value we already have and want to map into a new range
- `fromLow`: This is the lower limit of the possible input range
- `fromHigh`: This is the upper limit of the possible input range
- `toLow`: This is the lower limit of the possible output range
- `toHigh`: This is the upper limit of the possible output range

Finally, the function simply returns to us the mapped value corresponding to the output range.

Perhaps all this seems a little complex, and you don't want to make so many mathematical operations, but in just a moment, when we see our first example using analog inputs and outputs, you will immediately get the point and will love the map() function; believe me.

An ambient light meter

For our first example using analog inputs, I'd propose that you build an ambient light meter: a device that indicates the amount of ambient light it can sense from its environment in some way and shows it in a visual manner, which is a perfect beginning for your next burglar detector.

For the purpose of this example, we will use a very interesting electronic device called a photocell or light-dependent resistor. A photocell is simply a specific type of resistor that varies the resistance it offers according to the amount of light it receives from its environment, exhibiting a photoconductive behavior, that is, lowering its resistance as the light increases and vice versa, commonly ranging between a few ohms when exposed to a bright light and up to some mega ohms when totally in darkness.

In the following picture, you can see a typical photocell and the usual schematic symbols you can find to refer to it:

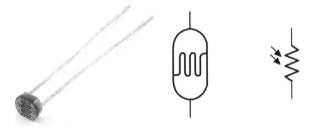

A real photocell and its most common schematic symbols

Unlike any other resistor, a photocell doesn't have a polarity, which means that you don't have to observe the way you connect it in your circuit. It does its job equally in one way or the other, but being a variable device it makes us give some considerations when connecting it to our Arduino board.

Connecting a variable resistor to Arduino

As mentioned previously, a photocell can decrease its internal resistance up to only a few ohms when exposed to a bright light and theoretically, it could even reduce it to zero when in the presence of a very bright light; in this case, it acts like just a simple wire with 0 ohms resistance.

With this in mind, we need to modify our circuit to prevent a short circuit in case this situation arrives.

Given that connecting variable resistors as analog input sensors to Arduino is a very common case, let's see a simple circuit that will ensure we will never damage the Arduino board.

A general variable resistor circuit to connect to Arduino can be like what is shown in the following diagram:

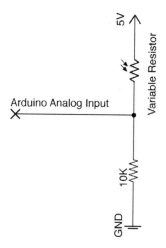

A typical circuit to connect a variable resistor to an Arduino analog input

As you can see, in some way, the circuit is very similar to the one used when connecting a digital input to Arduino. It also uses a 10K ohm load resistor to prevent a short circuit in case the photocell internal resistance decreases down to zero.

This is not a typical situation but it could happen, especially when connecting other types of variable resistors.

Voltage divider

This kind of circuit is commonly known in electronic jargon as a voltage divider, because it splits the provided voltage in two steps according to the proportion of the value of the first resistor to the other. We should keep this in mind when selecting the resistance we are going to use as protection.

We should always try to select a protection resistor of a value lower than the main variable resistor; this way, the main voltage drop will be placed just in the variable resistor acting as a sensor, that is, the photocell in our case.

This way, we allow Arduino to sense what is happening in the variable resistor with greater precision and use only the load resistor as protection.

In *Chapter 8, Communicating with Others*, we will revisit this special circuit to test and try different load resistor values, but for the moment, let's leave aside the theory and begin to work hands on with our next project: a device to measure the light surrounding Arduino.

An ambient light meter circuit

Here, you have my proposed circuit to be used as our first analog input example:

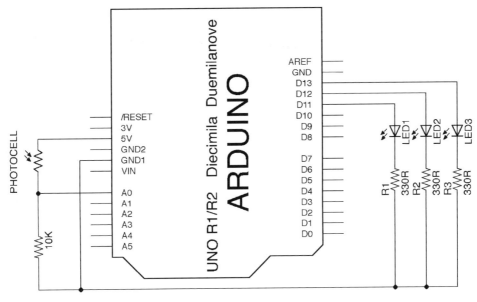

An ambient light meter circuit schematic

It has a voltage divider consisting of the photocell and a 10K ohm load resistor as the input circuit, just like the one I showed you in the previous section.

Regarding the output part of this circuit and trying to keep things simple, I have reused the assembly of the traffic light, but this time, we will use it as a level indicator with four possible states: low, medium, high, and maximum, being represented by none, one, two, or three turned-on LEDs.

From the point of view of used pins, I have connected the three LEDs to digital pins **11**, **12**, and **13** and the voltage divider to analog input **0**.

Breadboard connections

At this point in our work, I assume that you will be very accustomed to working with the breadboard, but in case you still have some doubts or the schematic is not clear enough for you, here you have the breadboard connection diagram of the ambient light meter project:

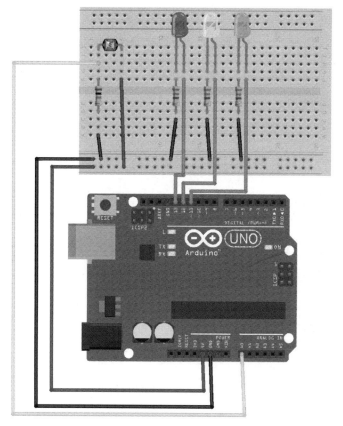

A breadboard connections diagram of the ambient light meter circuit

Programming to sense the light

From the point of view of programming, dealing with an analog input is not that different when compared to dealing with a digital one.

Just like when reading a digital input, we are going to receive a value from the reading function, analogRead(pin) in this case.

We will read from a pin that we have previously set as an input in the setup() function of our sketch by a call to the pinMode() function.

There is just one important difference. As we know from *Chapter 3, Interacting with the Environment the Digital Way*, the Arduino board has different pin headers for digital and analog input connections. When connecting something to your board that is going to be read in an analog way, you have to mandatorily use one of the **Analog In** pins of the Arduino board—those on the lower side of the board close to the **Power** pin headers.

You can only read an analog value through the use of the analogRead(pin) function from one of the **Analog In** pins of your board; definitely take this into consideration.

> Due to the fact that the analog inputs have their own exclusive pins, it is not compulsory to configure them as INPUT in the setup() function, because they are always inputs, but for the sake of clarity and learning purposes, I'll include this pinMode(pin, INPUT) function call in the examples.

An ambient light meter code

This example is a typical case that resolves a common situation when working with microcontrollers: it is common to have to action in a set of ways depending on an input value.

In our case, the project aims to turn on zero, one, two, or three LEDs depending on the magnitude of the ambient light surrounding the photocell.

Just for this typical case, most of today's programming languages have a similar control structure commonly known as switch / case.

The switch / case control structure

The general syntax of a `switch` / `case` control structure in the Arduino programming language is as follows:

```
switch (var) {
  case label:
    // statements
    break;
  case label:
    // statements
    break;
  default:
    // statements
}
```

The operation of this sentence allows us to opt for a branch of code execution depending on the value of a variable. In the preceding general syntax, we can end up executing any of the statements of the three case branches depending on the value of the `var` variable enclosed in parentheses following the `switch` word.

If none of the labels equal the variable value, the optional `default` section will be used, if present.

In our example, we will manage to obtain a variable that represents the light level divided into four possible states that will be used to separate the code in four different branches, each one turning on the LEDs accordingly.

Well, enough of theory for the moment. Let's take a look at the complete code of this project, and I'm sure you will get a clearer picture:

```
/*
 Chapter 06 - Analog Inputs to Feel Between All and Nothing
 Ambient light level
 By Francis Perea for Packt Publishing
*/

// Global variables we will use
// One for each pin we will use and
// two for the value reading and conversion
int redLED = 13;
int yellowLED = 12;
int greenLED = 11;
```

```
int photocell = 0;
int value = 0;
int state = 0;

// Configuration of the board: three outputs and one input
void setup() {
  pinMode(redLED, OUTPUT);
  pinMode(yellowLED, OUTPUT);
  pinMode(greenLED, OUTPUT);
  pinMode(photocell, INPUT); //optional
}

// Sketch execution loop
void loop(){
  // Read the sensor and convert the value to
  // one of the four states we will use
  value = analogRead(photocell);
  state = map(value,0, 200, 1, 4);
  // acts depending on the obtained states
  switch(state){
      case 1:
         digitalWrite(greenLED,LOW);
         digitalWrite(yellowLED,LOW);
         digitalWrite(redLED,LOW);
         break;
      case 2:
         digitalWrite(greenLED,HIGH);
         digitalWrite(yellowLED,LOW);
         digitalWrite(redLED,LOW);
         break;
      case 3:
         digitalWrite(greenLED,HIGH);
         digitalWrite(yellowLED,HIGH);
         digitalWrite(redLED,LOW);
         break;
      case 4:
         digitalWrite(greenLED,HIGH);
         digitalWrite(yellowLED,HIGH);
         digitalWrite(redLED,HIGH);
         break;
   }
 }
```

The potentiometer

Two typical potentiometers along with their corresponding schematic symbol are shown as follows:

Two typical potentiometers and their schematic symbol

The component itself has three legs. If you measure the resistance between the external two legs, you should get the total resistance of the potentiometer, but if you take the measure between the central leg, usually called cursor, and any other, you will get a resistance proportional to the rotational angle of the potentiometer.

In our project, we will use it as a kind of a throttle for our DC motor by making the speed of the motor a function, or dependent, of the potentiometer position.

The motor speed control schematic

Here, you have the project schematic. It's nothing new, as you can see. The input side of the circuitry is similar to the input part of the ambient light meter, and the output side is similar to that of the motor driver in Chapter 4, *Controlling Outputs Softly with Analog Outputs*.

I have once again used pin **A0** to read the input and in this case, I will act on PWM pin **6** to control the motor speed though a transistor.

Chapter 6

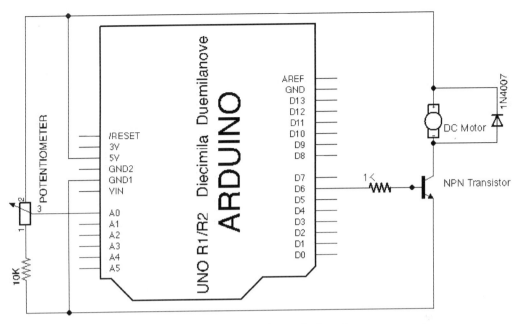

The DC motor speed control schematic

The breadboard connections diagram

Next, you can find the breadboard connections diagram for the proposed circuit.

The only components you should care about are the diode and the transistor.

The diode should be correctly placed because of its polarity. The bar indicating its cathode should be in contact with 5V.

Regarding the transistor, I used a BC547 with its legs ordered as CBE in my
assembly; keep this in mind or refer to your transistor datasheet to know its
own pin out before connecting it.

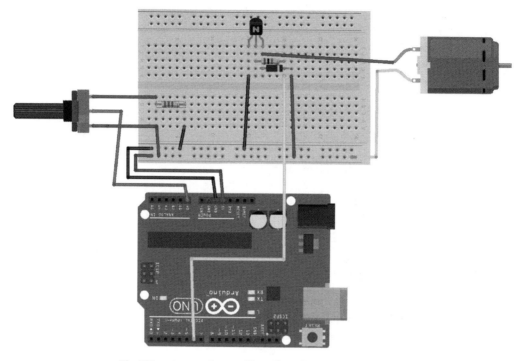

The DC motor speed control breadboard connections diagram

A simple code to control the motor speed

The code for this project couldn't be simpler. Just read, convert, and output the
mapped value to the motor. Here is the code:

```
/*
 Chapter 06 - Analog Inputs to Feel Between All and Nothing
 DC Motor speed control
 By Francis Perea for Packt Publishing
*/

// Global variables we will use
```

```
// One for for the potentiometer
// and another to command the transistor
// Two variables to store read and converted values
int pot = 0;
int base = 6;
int potvalue = 0;
int motorspeed = 0;

// Configuration of the board: three outputs and one input
void setup() {
  pinMode(base, OUTPUT);
  pinMode(pot, INPUT); //optional
}

// Sketch execution loop
void loop(){
  // Read the sensor and convert to the
  // allowed output range for an analog output
  potvalue = analogRead(pot);
  motorspeed = map(potvalue, 512, 1023, 100, 255);
  analogWrite(base, motorspeed);
}
```

Once again, I have used a little bit of calibration to get to the best values for the input and output ranges of the `map()` function.

By using a 10K ohm potentiometer and a 10K ohm protection resistor, I got values from `512` to `1024` from the potentiometer.

Regarding the motor, outputting less than `100` was unable to make it spin, so finally, my conversion is `motorspeed = map(potvalue, 512, 1023, 100, 255);`.

As I told you previously, if you can't wait to know how to calibrate your sensor readings, you can take a look at *Chapter 8, Communicating with Others*.

Summary

This was a well-packed chapter. We have learned a lot of concepts in this chapter, both in terms of software and hardware.

From the point of view of concepts, we came to know about ADCs, their resolution, and the problem they bring when having to change values from one range to another. The `map()` function came to our help.

Regarding the physical side, we met two very practical electronic components: the photocell and the potentiometer. Both will help us when developing our projects. We also learned about what a voltage divider is and how it can help us when connecting external sensors to our Arduino board.

Finally, on the programming side, we saw how to read from an analog input with the `analogRead()` function, and we learned about the use of the powerful `switch / case` function to allow us to take different ways of execution.

Arduino still has some features that deserve a play, so let's go ahead to the next chapter where we will try to conquer the domain of time.

7
Managing the Time Domain

We have come a long way, and at this point we have seen almost everything regarding digital and analog inputs and outputs. But Arduino still has some unseen characteristics that come in very handy when developing our projects. Controlling the time is one of them and a very important feature for a good number of projects.

In this chapter, we will learn about the functions in the Arduino library to control the time and will also discover sound generation and the use of speakers and buzzers. Every timer has an alarm, doesn't it?

Time control functions

The Arduino library has four functions that allow us to manage the time. We have already seen one of them, the `delay()` function, which we have used from our first sketch to stop the execution of the code for a short period of time.

As we saw in *Chapter 3, Interacting with the Environment the Digital Way*, the `delay()` function accepts only one parameter: the desired number of milliseconds to pause. For most projects, this resolution will be fine, but there can be situations where a millisecond is too much.

For these kinds of problems, the Arduino library also offers the `delayMicroseconds()` function that, as you can imagine, pauses the code execution for just the number of microseconds that you set as its only parameter.

A microsecond is a millionth of a second, or to put it another way, there are a thousand microseconds in a millisecond and a million microseconds in a second.

Due to restrictions in the Arduino architecture, the maximum delay the `delayMicroseconds()` function can produce is around 16,000 ms. In case you should need a larger delay, you can always use your old friend, the `delay()` function.

Stopping versus accounting

Both functions that we already know, `delay()` and `delayMicroseconds()`, stop the code execution for a desired amount of time, but we don't always want to stop. In some cases, we just want to know what time is it, or if a certain amount of time has passed by.

For accounting purposes, the Arduino library offers two other functions that don't stop the code, but simply return a value representing a time. We can use this to store in a variable to operate with it by making calculations or by taking decisions based on this time.

These two functions are as follows:

- `millis()`: It returns us the number of milliseconds since the sketch execution began
- `micros()`: It returns the number of microseconds elapsed from the beginning of the sketch execution

Both return a value of type `unsigned long`, which means that they use 4 bytes (32 bits) to store it and that the maximum value they can store is 4,294,967,295 or 2^32 - 1.

Due to this maximum value, if we keep our sketch running for a very long time, the value returned by these functions may overflow, that is, may return to 0 and start counting again.

In the case of the `millis()` function, the overflow will happen after approximately 50 days and in the case of the `micros()` function around 70 minutes after the beginning of the sketch execution, so take this into consideration when using these functions in long runner projects.

Making some noise

We usually associate time with sound, from the tick-tock of an old analog clock to the sound of an alarm in our latest digital clock.

From the hardware point of view, in this chapter I'll take advantage of this association to show you how to connect a speaker or a buzzer to our Arduino board and the way Arduino can generate sounds.

Being able to produce some sounds allows us to account for time in a more sensorial way.

Arduino library sound functions

Because it is a very common task, once again the Arduino library comes to our help with the `tone()` function that will help us produce those electrical signals that we need to generate different sounds.

The `tone()` function accepts up to three parameters:

- `pin`: The pin number through which we want to generate the sound signal
- `frequency`: The frequency in hertz at which we want the signal to oscillate
- `duration`: An optional parameter that specifies the total duration of the sound

Everything is almost as expected, but perhaps the `duration` parameter needs a little explanation.

The `tone()` function permits two ways of calling it, with or without the `duration` parameter. If you specify the `duration` parameter, the sound will be generated just for that time; once the time has passed by, the sound stops.

In its other format, that is, without the `duration` parameter, it's your responsibility to stop the sound by calling the `tone()` counterpart function, `noTone()`, which takes the pin as its only parameter where you want the sound to stop.

You should also know that Arduino can only generate a sound at a time through the same pin. This means that if you call the `tone()` function on a pin that is already generating a sound, the former will be stopped and the latter one will be produced.

Let's see this more clearly with a pair of examples.

Suppose we included a code like the following in our sketch:

```
...
tone(pin, 440, duration);
tone(pin, 220, duration);
...
```

We will simply hear the second tone, the one of 220 Hz, because despite having set a `duration` parameter, we execute just another `tone()` call in the next sentence, which will directly override the previous tone generation.

The `duration` parameter will make the tone last for the specified time provided that we don't produce another tone in the same pin before reaching this duration.

The correct way of generating more than one sound in a sequence could be something like the following code:

```
...
tone(pin, note1, duration);
delay(duration)
tone(pin, note2, duration);
delay(duration);
...
```

The other way of generating more than one sound is directly without the use of the `duration` parameter and turning the tone off after the desired duration by calling the `noTone()` function:

```
...
tone(pin, note1);
delay(duration)
noTone(pin);
tone(pin, note2);
delay(duration);
noTone(pin);
...
```

This way, we ensure that no other tone is being generated until the previous one has been finished.

The major disadvantage of programming sound generation this way is that you can't do anything else while a sound is being produced. In the next project, I'll show you a more elaborate technique that uses the `millis()` function and allows you to continue working while waiting for the note to finish.

Sound hardware connection

The next thing we should look for when trying to generate a sound is a speaker or a buzzer, a device that when electrically excited can produce a sound that varies with the frequency of the electrical signal we use.

To connect a speaker to Arduino, the only thing you need is a free pin through which we can generate an electrical signal of a specific frequency by calling the `tone()` function. Even so, there are two different ways to connect the speaker to Arduino:

- Directly
- Through a transistor that acts as a driver

Let's see both cases along with their advantages and disadvantages.

Direct connection

In case you are using a small speaker, you could drive it directly from one Arduino pin without the need of additional circuitry as shown in the following schematic:

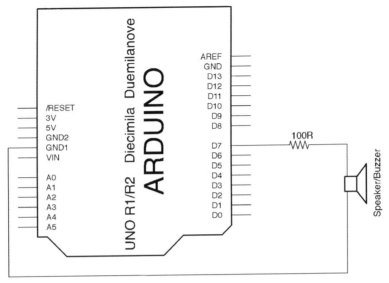

Direct connection of a speaker to Arduino

Clearly, the main advantage of this way of connecting the speaker is its simplicity. Just a small 100 Ω resistor is all you need to make it sound.

The biggest disadvantage of this way of connecting the speaker is that, as we have talked about earlier, Arduino can only deliver up to 40 mA current per pin, and with such a small current, a typical speaker won't produce an adequate sound.

If your project simply needs a way to give small sound signals, then this is your circuit, but if you need a powerful alarm with a sound loud enough to be heard several meters away, you should definitely opt for the transistor driver connection that we will see in the next section.

Connection through a transistor

A speaker is no more than a coil around a magnet, that is, an inductive load. We have already dealt with other inductive loads in other projects, motors to be precise, in *Chapter 4, Controlling Outputs Softly with Analog Outputs*.

Just like when dealing with motors, we will use a transistor as a device to regulate the current that flows through our load, the speaker in this case. Big speakers consuming bigger currents can produce considerable back electromotive spikes when powered off, just like motors, and to prevent them we will use a diode as you can see in the following circuit schematic:

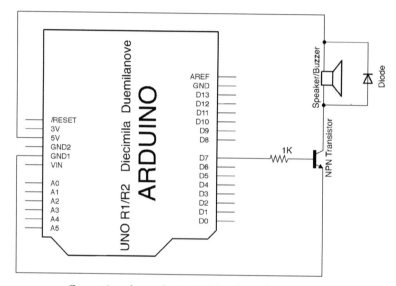

Connection of a speaker to Arduino through a transistor

Compared with the previous circuit, the main advantage of this way of connecting the speaker is its output power. The sound we can produce with such a circuit is much louder than with the previous direct connection.

Obviously, it is a slightly more complicated circuit to assemble, but not too much.

In case you still aren't able to assemble this circuit on your breadboard, here you have the breadboard connections diagram for the preceding schematic using a transistor with an EBC pinout as shown in the following image:

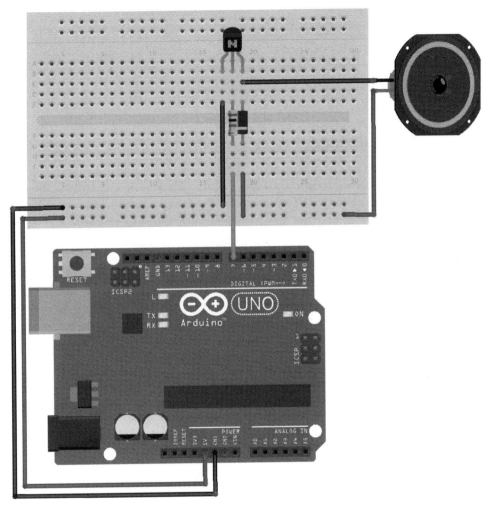

Breadboard connections diagram to connect a speaker to Arduino through a transistor

To make use of all of the concepts that we have learned about so far, I would like to propose to you a project that manages time and produces some sounds.

Managing the Time Domain

A simple timer

We are going to build a timer that once it reaches a predefined amount of time, generates a simple chord, repeating this behavior three times.

First, I'll show you a sketch that uses the `delay()` function to wait for the time lapse to pass by and also to wait for every note to be played. Obviously, with this technique we can't do anything but wait in the time between chords.

In a more advanced example, we will see another sketch that uses the `millis()` function to accomplish just the same task but this will allow us to do other things while waiting for the time to pass by, blinking an LED in this case.

Dividing your sketch into different files

I will also use this code example to introduce you to the possibility of including other files in our code, so that we can create modular sketches dividing the sketch into various files.

In this case, I've created a file with the definition of every musical note frequency, so that I can call the `tone()` function with the note name directly, instead of using a frequency.

You have to create a new tab in your sketch by using the down-pointing arrow at the right side of the current tab name and selecting **New Tab** as shown in the following screenshot:

New Tab command

You can find the complete file in the accompanying source code, but for the sake of clarity, its contents are as follows:

```
#define NOTE_B0   31
#define NOTE_C1   33
#define NOTE_CS1  35
#define NOTE_D1   37
#define NOTE_DS1  39
#define NOTE_E1   41
...
```

As you can see, we are just defining a series of replacements of the note names for their frequency and the Arduino programming environment will substitute them just before compiling our code. This makes the sketch much more easily readable and maintainable.

I have called this file `pitches.h` and I have to include this in the main file by simply using a line of code like this:

```
#include "pitches.h"
```

Coding a timer by using delays

Here, you have the complete code for the main file of this first sketch for our timer as follows:

```
/*
 Chapter 07 - Managing Time Domain
 Timer with delays
 By Francis Perea for Packt Publishing
*/
// Load the notes frequencies definitions
#include "pitches.h"

// Set some global constants
// Total amount of time to wait
#define lapse 3000
// Time to play every note of the chord
#define noteDuration 500

// Global variables we will use
// Speaker pin
int buzzer = 7;
// The number of chords played
int numberOfSounds = 0;

// Configuration of the board: just one output
void setup() {
  pinMode(buzzer, OUTPUT);
}

// Sketch execution loop
```

```
void loop(){

  // If less than 3 chords played
  if (numberOfSounds < 3) {
    // Wait for the desired lapse
    delay(lapse);
    // Play every notes and wait
    tone(buzzer, NOTE_G4, noteDuration);
    delay(noteDuration);
    tone(buzzer, NOTE_A5, noteDuration);
    delay(noteDuration);
    tone(buzzer, NOTE_C5, noteDuration);
    delay(noteDuration);
    tone(buzzer, NOTE_D5, noteDuration);
    delay(noteDuration);

    // Incremet the number of played chords
    numberOfSounds += 1;
  }
}
```

At this stage of our journey, I'm almost sure you can understand easily what this sketch does, but let's take a general look.

We just maintain the `numberOfSounds` variable with the number of chords played and in the case it is less than three, we simply wait for the desired lapse and play every note in sequence waiting for each one to finish.

Notice that the comparison in the `if` sentence says `numberOfSounds < 3`, and not `numberOfSounds = 3`, because in the first loop, the `numberOfSounds` variable will have a value of 0 and if we wait for it to reach a value of 3, we will do four iterations of the loop instead of the desired three.

I have also included a little facility in the code. To increment the `numberOfSounds` variable just after playing the chord, I have used the sentence `numberOfSounds += 1`, which is simply an abbreviation of `numberOfSounds = numberOfSounds + 1`, a very common construction in C programming.

Coding without delays and blinking an LED while waiting

To show you how to work with the `millis()` function, I would like to show you another sketch to accomplish just for the same task of the previous section but without the need to stop code execution while waiting, so that we can do other things, like blinking an LED to reflect time passing by.

The circuit is very similar to the previous one but just includes an LED to blink. Here, you have the circuit schematic and the breadboard connections diagram:

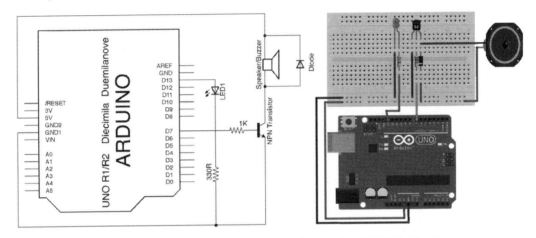

Circuit schematic and breadboard connections diagram for an LED blinking timer

You have the complete code for this example as follows, which is a little more complex than the previous one, but in a moment you will understand it:

```
/*
 Chapter 07 - Managing Time Domain
 Timer without delay
 By Francis Perea for Packt Publishing
*/

// Load the notes frequencies definitions
#include "pitches.h"

// Set some global constants
```

Managing the Time Domain

```
// Total amount of time to wait
#define lapse 3000
// Time to play every note of the chord
#define noteDuration 300
// Time to blink the LED
#define blinkDuration 500

// Global variables we will use
// Speaker pin
int buzzer = 7;
// The number of chords played
int numberOfSounds = 0;
// LED pin
int LED = 13;
// Initial LED staus
int LEDStatus = LOW;
// The time when the LED blinked last time
unsigned long blinkMark;

// Configuration of the board: two outputs
void setup() {
  pinMode(buzzer, OUTPUT);
  pinMode(LED, OUTPUT);
  // Set the LED initially off
  digitalWrite(LED, LEDStatus);
  // Record the first blink
  blinkMark = millis();
}

// Sketch execution loop
void loop(){
  // If less than 3 chords played
  if (numberOfSounds < 3) {
    // Wait for the desired lapse but blink in the meantime
    myActiveDelay(lapse);

    // Play every note and wait blinking
    tone(buzzer, NOTE_G4, noteDuration);
```

```
    myActiveDelay(noteDuration);

    tone(buzzer, NOTE_A5, noteDuration);
    myActiveDelay(noteDuration);

    tone(buzzer, NOTE_C5, noteDuration);
    myActiveDelay(noteDuration);

    tone(buzzer, NOTE_D5, noteDuration);
    myActiveDelay(noteDuration);

    // Incremet the number of played chords
    numberOfSounds += 1;
  }
}

void myActiveDelay(int timeLapse){
 // Record the time
 unsigned long delayMark = millis();
 // wait for the desired timelapse to reach doing something
 while (millis() - delayMark < timeLapse){
    tryToBlinkaLED();
    }
 }

void tryToBlinkaLED(){
    // if the blinking duration has arrived
    if (millis() - blinkMark > blinkDuration){
      // invert the LED status
      LEDStatus = !LEDStatus;
      // turn the LED accordingly
      digitalWrite(LED,LEDStatus);
      // record new blink time
      blinkMark = millis();
    }
}
```

The first thing I'd like to note is that I have divided the code into different functions that I've created to make the complete code more easily understandable. I have created the following two functions:

- `myActiveDelay()`: This function just waits for the time to pass by and in the meantime makes other things; in our example, it simply calls the function that blinks the LED. We store the value of `millis()` in the `delayMark` variable just at the beginning of the function and wait for `millis()` to be bigger than this stored value in just the desired time lapse. To wait for the time to pass by, we use the `while` control structure that executes the instructions inside as long as the condition inside the parentheses is true. You can learn more about this kind of loop in the while page at the Arduino site reference section at http://arduino.cc/en/Reference/While. Also, please note that the variable I have used to store the time mark is of the type `unsigned long`, just like the returned value of the `millis()` function.

- `tryToBlinkaLED()`: In a very similar way to the previous function, it simply checks if the blinking time has passed by and in that case changes the LED status and records the new blinking time. To record the status of the LED, we use a global variable called `LEDStatus` and inside this function, we commute its value from `LOW` to `HIGH` and vice versa by using a very common C language construction, `LEDStatus = !LEDStatus`, that makes `LEDStatus` pass from `0` to `1` and from `1` to `0`, or from `HIGH` to `LOW` and from `LOW` to `HIGH` in the same way.

Once you understand the mission of these two functions, you could see that the main loop is practically identical to the one of the previous examples. We simply check if less than three chords have been played, in which case we play every note in sequence and wait for them to finish in an active manner, which is blinking an LED.

Perhaps, you should play a little with this code, and try to replace the `tryToBlinkaLED()` function by any other function you consider of interest or change the lapses and durations to see their influence in the final result.

A bigger project – a metronome

Being a saxophonist myself, one of the tools I use the most when practicing is my metronome. A metronome is a device that gives pulses at a selectable frequency and that helps us maintain the correct tempo when playing music.

The metronome has a double indication; it produces clearly audible ticks and also swings a pendulum to give us visual feedback.

With the next project, I'd like to propose that you build a digital metronome with Arduino. If you think about it carefully, you will notice that we have almost created the core of the metronome in the previous example.

We will only add a pair of buttons to allow us to change the tempo and another LED to simulate the pendulum swing by oscillating between two LEDs.

Let's go on and try to finally conquer the time domain.

The metronome circuit

The circuit of this project is a bit more complicated than any other so far, because it has to deal with up to five peripherals: two LEDs, two buttons, and a transistor to drive the speaker, but if you take a look at it carefully and take only a device at a time, it is not much more difficult to understand than any of the previous circuits in this book.

Here, you have both the circuit schematic and the breadboard connections diagram shown, which will also help you to understand the complete project:

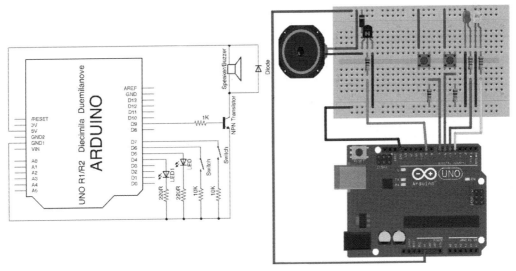

Circuit schematic and breadboard connections diagram for a digital metronome

The metronome code

The complete code for the proposed metronome project is as follows:

```
/*
 Chapter 07 - Managing Time Domain
 Metronome
 By Francis Perea for Packt Publishing
*/

// Load the notes frequencies definitions
#include "pitches.h"

// Define constants we will use
#define buzzer 9
#define decButton 7
#define incButton 6
#define redLED 5
#define greenLED 4
#define tickDuration 100
#define minBPM 50
#define maxBPM 220
#define pressIncrement 10

// Global variables we will use
// The initial tempo
int bpm = 60;
int beatDuration;
boolean readButtons;

// Configuration of the board
void setup() {
  pinMode(buzzer, OUTPUT);
  pinMode(decButton, INPUT);
  pinMode(incButton, INPUT);
  pinMode(redLED, OUTPUT);
  pinMode(greenLED, OUTPUT);
}

// Sketch execution loop
void loop(){
  // Allow buttons to be read
  readButtons = true;
```

```
  // Compute new beat duration
  beatDuration = ((float)60 / bpm) * 1000;
  // Make beat
  beat();
}

void beat(){
   // Turn LED on
   digitalWrite(redLED, HIGH);
   // Generate half beat sound
   tone(buzzer, NOTE_C3, tickDuration);
   // Wait for sound to finish while looking at the buttons
   myActiveDelay(beatDuration);
   // Turn LED off
   digitalWrite(redLED, LOW);

   //Repeat for the other beat half
   digitalWrite(greenLED, HIGH);
   tone(buzzer, NOTE_B4, tickDuration);
   myActiveDelay(beatDuration);
   digitalWrite(greenLED, LOW);
}

void myActiveDelay(int timeLapse){
 // Record the time
 unsigned long delayMark = millis();
 // wait for the desired timelapse to reach
 while (millis() - delayMark < timeLapse){
   //and look at the buttons in the meantime
   if (readButtons == true){
     checkButtons();
   }
 }
}

void checkButtons(){
   // If haven't reached minimum tempo and decrement button pressed
   if ((digitalRead(decButton) == HIGH) && (bpm > minBPM)){
    // Decrement tempo
    bpm -= pressIncrement;
    // Disable buttons so they are not read again until next beat
    readButtons = false;
```

```
        }
        //The same for increment button and maximum tempo
        if ((digitalRead(incButton) == HIGH) && (bpm < maxBPM)){
          bpm += pressIncrement;
          readButtons = false;
        }
    }
```

There are some small considerations to be taken into account in respect of this code.

You will notice that I have used definitions instead of declaring variables. Given that these values won't change during the code execution, we can substitute them for `#define` instructions that don't consume any memory space.

Regarding the variables, I have declared three:

- `bpm`: This will store the current tempo and will allow us to modify it when the corresponding button is pressed.
- `beatDuration`: This will simply help us make the needed calculation to convert from **beats per minute** (**bpm**) to the corresponding duration of a beat.
- `readButtons`: This variable is what is usually called a flag, so I have declared it as a `boolean`, a type of variable that will only store values of `true` and `false`. I will use it to prevent multiple pressings of the buttons, and limiting to just one press in a beat. It is set to `false` once a button has been pressed and again to `true` in the next beat. This way, we only check for button pressings while waiting for the time to pass by when this variable is `true`.

There is also an important consideration regarding variable types and operations. You may have noticed the line where I compute the `beatDuration` value in every main loop iteration as follows:

```
    beatDuration = ((float)60 / bpm) * 1000;
```

The (`float`) before `60` is to force the result of the operation to be a floating point number. If you don't include it, the result of, for example, `60 / 65` will be simply `0` and not `0.92`.

The rest of the code should be easily understood simply by following the comments, as it has quite a lot of common code with the previous examples.

Don't hesitate to assemble the circuit and play with the code, by changing some of the defined constants and even trying to enhance the functionality of the project.

Summary

We have seen lots of concepts related to not only time in this chapter, but also some hardware and programming concepts.

Mainly, we have learned about the different time functions the Arduino library gives us to use in our sketches and the difference between waiting with `delay()` and `delayMicroseconds()` and accounting with `millis()` and `micros()`.

From the hardware point of view, we have connected a speaker to Arduino in two different ways: directly and through a transistor.

To use a speaker, we saw the use of the `tone()` and `noTone()` functions and their features.

Talking about programming, we learned to divide our code into different files and into different functions inside a file, which is a very common technique that will help you construct modular code.

We even met the `while` control structure, a new kind of conditional loop that evaluates a condition before executing the corresponding instructions.

Perhaps, you think that there can't be much more to learn from such a small board, but in the next chapter, I'll show you one of the most used facilities in the Arduino board, its **Universal Asynchronous Receiver-Transmitter** (**UART**), which allows Arduino to communicate with a wide number of devices, including your own computer.

Let's go ahead and open a door that will lead you to a brand new level and that will enable you to create projects that speak to others.

8
Communicating with Others

Until now, all our projects have been designed to be standalone and independent, but there will be occasions where we will have to integrate our project with other external devices, and this means that we will have to make our Arduino talk with these other elements. Usually, this communication is made via a very practical and extended standard called serial communication.

In this chapter, we will see how to connect and program Arduino so that it is capable of serially talking to our computer as the simplest and most common type of serial communication, but what you will learn here is directly applicable to any other connection you have to establish to another device that can talk serially.

So, let's give up talking and let's make our projects talk.

Serial communications concepts

All through this chapter, when we talk about serial communication, we will be referring to RS-232 standard protocol communication based on the **Universal Asynchronous Receiver/Transmitter** (**UART**) that the Arduino microcontroller incorporates. It is the most common communication type for most Arduino projects in comparison to other serial protocol communications that can also be established with Arduino and that we will briefly introduce in the next section.

The first thing we should know is why all these communication systems are called serial as opposed to parallel communications.

In the case of a serial communication, every bit of the data being transmitted is sent one piece at a time, through just a single line of the communication channel. In a parallel communication, data bits are sent in groups all at once, which makes it necessary to have a bigger number of communication lines in the channel.

This is perhaps the main reason to use serial communications in Arduino, due to the limited number of pins available to establish the communication channel, allowing using just two pins in the case of RS-232 communications.

If you take a closer look at your Arduino board, you should notice a pair of pins marked as **TX** and **RX** in the digital pins row, pins **0** and **1** to be precise, as you can also see in the following image:

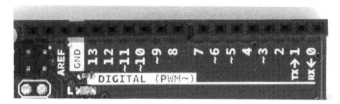

TX and RX pins in the digital pins row

When connecting Arduino to your computer, you don't really have to use these pins; instead, you can establish a serial connection through the USB port thanks to the UNO ATmega16U2 chip, or the FTDI chip in older boards, which tunnels the serial communication over the USB.

In case you want to connect your Arduino to any other kind of serially capable device, you should use the TX/RX pins instead.

This facility requires a little consideration when connecting external circuitry to your Arduino board through these pins. Since they are going to be used even when uploading your sketches to the microcontroller, you may have to disconnect what you have attached to these pins when uploading the code and connect them back again once the communication has finished.

My personal advice is to avoid using these pins as long as you have others available, and use them only if you don't have any more free or exclusively available when developing projects in which you have to establish serial communication with other serial devices.

All these pin considerations apart, I won't go much deeper into the technical features of a serial communication, precisely because of its implicit simplicity, but there is just one specific parameter relating to serial communications that is necessary to know: the baud rate.

The baud rate

The baud rate in a serial communication specifies the data rate in bits per second that both communicating devices must comply with to be able to understand each other. Typical values range from 300 to 115,200, depending on the connected devices.

There are also other parameters to be specified that affect the communication but are optional, like the number of data bits, parity, and stop bits.

If you are interested in the internal mechanisms that regulate a serial communication, you could visit the Wikipedia page for the serial communication at http://en.wikipedia.org/wiki/Serial_communication and from the point of view of Arduino, you could also take a look at the **Serial** part of the **Reference** section of the Arduino site at http://arduino.cc/en/Reference/Serial.

Other types of serial communication

Arduino does not only support RS-232 serial communication, but also supports two more communication protocols:

- **Serial Peripheral Interface (SPI)**: A special master/slave serial communication protocol used in short distance communications and very popular among different types of sensors. Its main disadvantage is the need for four pins to establish the communication channel.
- **Inter-Integrated Circuit (I2C)**: A bus-based master/slave communication protocol allowing for multimasters and multislaves. It is mainly aimed at connecting low speed devices.

They are out of the scope of this book and I will simply give you some links in case you want to investigate a little more by yourself:

- Regarding SPI, you can read its Wikipedia page at http://en.wikipedia.org/wiki/Serial_Peripheral_Interface_Bus
- To know how to deal with SPI from Arduino, there is a special dedicated page to this protocol at the Arduino **Reference** section of the Arduino site at http://arduino.cc/en/Reference/SPI
- If you are interested in connecting your Arduino to an I2C bus, you should first read the I2C page of the Wikipedia website at http://en.wikipedia.org/wiki/I²C

- And to learn how to program I2C, you could visit the Arduino **Reference** section and read about the Wire library at `http://arduino.cc/en/reference/wire`

Arduino easily supports both of these protocols by using dedicated libraries that come included in the Arduino programming environment. The use of external libraries is a topic that we will see in a later section of this chapter.

Calibrating sensors serially

In *Chapter 6, Analog Inputs to Feel between All and Nothing*, I promised you that I'd show you how we can use a serial communication to calibrate the analog sensors you could connect to your projects, in particular in the ambient light meter and the motor speed control projects.

The time has now come to revisit these projects and finally unveil how we got the correct output range for the mapping we made in these projects.

If you remember, we used the following circuit for the ambient light meter as shown in the following image:

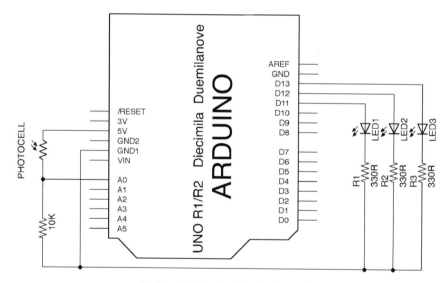

Ambient light meter circuit schematic

In the sketch, we had to read the photocell value and map it to a four-state output range with the following lines of code:

```
...
void loop(){
  // Read the sensor and convert the value to
  // one of the four states we will use
  value = analogRead(photocell);
  state = map(value,0, 200, 1, 4);
  // acts depending on the obtained states
  switch(state){
      case 1:
...
```

At that time, I didn't tell you how I got that input range of 0–200, but now I will cover this.

By using just the input part of this circuit, the photocell, and the 10K Ohm protecting resistor, we are going to use the following sketch to be introduced to serial communication with Arduino and read the values generated by the photocell so that we can get an idea of the possible range of values to expect from it:

```
/*
 Chapter 08 - Communicating with others
 Serial sensor reading
 By Francis Perea for Packt Publishing
*/

// Photocell pin as a define doesn't consume memory
#define photocell 0

// Global variables we will use
// A variable to store the read value
int value = 0;
// A variable to convert the read value to our output range
int state = 0;

// Configuration of the board: just one output
void setup() {
```

```
    pinMode(photocell, INPUT); //optional
    // Init serial communication
    Serial.begin(9600);
}

// Sketch execution loop
void loop(){
    // Read the sensor and convert the value to
    // one of the four states we will have
    value = analogRead(photocell);
    state = map(value,300, 550, 1, 4);
    // Send the read value and converted state through
    // the serial communication in a fancy way
    Serial.print(value);
    Serial.print(" : ");
    Serial.println(state);
}
```

The code is quite simple, and as you should have noticed it only reads the sensor value, and maps it to a new range. However, it also includes all necessary instructions to establish and maintain a serial communication with the computer to send the read and converted values to it so that you can know them and adapt your input range in response.

To begin, in the `setup()` function, we simply establish a serial communication at 9,600 bauds by using the following line of code:

```
Serial.begin(9600);
```

At this moment, it really doesn't matter to you, but to establish a serial communication we are going to use the `Serial` object, and thus we have to use the `object.method` syntax, `Serial.begin()` in our case.

For our purposes in this chapter, all we have to know is that with this instruction we begin the serial protocol to make the Arduino board communicate with any serial device, the computer in our example.

Once the serial communication is established, we have to just send whatever value we want through it to the other part. To make this in the previous code, we have used three different lines of code:

```
Serial.print(value);
Serial.print(" : ");
Serial.println(state);
```

The only difference between the `Serial.print()` and `Serial.println()` methods is that the former simply sends the value enclosed in the parentheses, while the latter sends the value and a carriage return, which comes in very handy when trying to format the serial output.

In our case, I've even included a delimiter between the read value and the converted one, so that the final result will be something like `508 : 3`.

To see it in action, all you have to do is upload the code to the Arduino board and open, as we saw in *Chapter 2, The Arduino Development Environment*, the Serial Monitor from the last icon on the Arduino programming environment toolbar, the one at the top right separated from the rest and whose icon represents a loupe looking at bits as shown in the following image:

The Serial Monitor icon at the right most part of the Arduino toolbar

Once the window opens, the first thing you have to look at is that the select baud rate option at the bottom right of the window is the same as that you have set in the code the Arduino runs. This ensures that both the Arduino and your computer are talking at just the same speed. In our case, we set the Arduino to begin the communication at 9,600 bauds, so your window should look similar to the one shown in the following image:

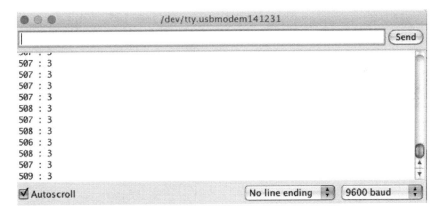

The Serial Monitor window showing incoming data at 9,600 bauds

Once the correct baud rate is set, you will begin to see a constant flow of data appearing on the window. In our case, the data will represent the read value from the sensor and the corresponding mapped value separated by a colon.

If data is not constantly appearing, ensure that you have checked the **Autoscroll** checkbox that allows the window to always scroll down to the new incoming data automatically.

In the next section, we will see how to use the upper textbox to send data from the computer to Arduino.

Once you are able to read the data flow the Arduino is sending to your computer, you can cover your photocell to see how this data changes immediately to reflect the ambient light variation.

If you take the photocell to its two extreme situations, totally covered and highly illuminated, you can take note of the minimum and maximum values the sensor gives and these values, as you will have supposed, correspond to the minimum and maximum values of our input range for the mapping into the new output range.

In this example, I got values from 300 to 550, but they depend on my particular photocell, the place where I placed the photocell, and even the time of day, so take note of your readings and adapt your code in response.

In most of the projects, once the calibration has been made you simply comment out the lines corresponding to the serial communication to disable them and not make Arduino work on it if you're not going to need them anymore. Don't permanently delete them in case you need to later take a new look at the sensor readings.

Sending data to Arduino

What we have seen till now is just one half of a serial communication; we have simply sent data from the Arduino to our computer. In this new example, we will revisit the motor speed control project of *Chapter 6, Analog Inputs to Feel between All and Nothing*, and replace the potentiometer by our computer in the sense to use it as a way to vary the motor speed, and take advantage of this new control method to incorporate two new possibilities:

- Totally stop the motor
- Make it run at full throttle

Since we are not going to use the input side of the previously mentioned project, we only need to connect a motor to Arduino as we already saw in *Chapter 4, Controlling Outputs Softly with Analog Outputs*, and I'll include that schematic here again for better understanding:

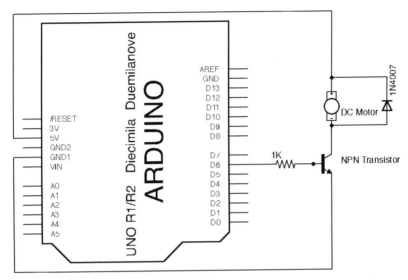

Connection of a motor to Arduino to be controlled via a serial communication

The complete code of the sketch we are going to use for this data-sending example is as follows:

```
/*
  Chapter 08 - Communicating with others
  Sending data to Arduino
  By Francis Perea for Packt Publishing
*/

// Global Definitions
// The pin used for the transistor base
#define transistorBase 6
// The increment for the motor speed
#define motorIncrement 10

// Global variables we will use
// A variable to store the received value
int dataReceived;
```

```
// and another to set the motor speed
int speed = 0;

// Configuration of the board: three outputs and one input
void setup() {
  pinMode(transistorBase, OUTPUT);
  // Init serial communication
  Serial.begin(9600);
}

// Sketch execution loop
void loop(){
  // check if data has been sent from the computer:
  if (Serial.available()) {
    // read the next char
     dataReceived = Serial.read();
    // Act depending on it
    switch (dataReceived){
      // Increment speed
      case '+':
        if (speed<250) {
          speed += motorIncrement;
        }
        break;
      // Decrement speed
      case '-':
        if (speed>5) {
          speed -= motorIncrement;
        }
        break;
      // Stop motor
      case '0':
        speed = 0;
        break;
      // Full throttle
      case '*':
        speed = 255;
        break;
   }
   // Send back the actual speed
   Serial.println(speed);
```

```
    // Set the speed motor
    analogWrite(transistorBase, speed);
  }
}
```

To begin, we have set two definitions, `transistorBase` for the pin that we'll use to connect the transistor base and `motorIncrement` to specify the increment or decrement we'll use when changing the motor speed.

Once done, we declared two variables, `dataReceived` to hold the data received through the serial communication and `speed` to account for the current motor speed.

In the `setup()` function, we simply set the transistor base pin as an output and began a serial communication at 9,600 bauds.

The main loop is where all the action is taking place and where the interesting part regarding serial data sending occurs. We just look if any new data has come through the serial communication with a call to the `Serial.available()` method, which returns `true` in case new data has been sent.

If we have new data, we get it 1 byte at a time by storing it in the `dataReceived` variable with a call to the `Serial.read()` method, and based on the value of this variable, we set a `switch` control structure to act depending on the received character.

The code considers four different situations:

- Increment of the motor speed when it receives a + character
- Decrement of the motor speed when it receives a - character
- Full halt of the motor if a 0 is received
- Set the motor at full throttle in case a * is received

In every case, we simply change the value of the `speed` variable accordingly.

Just after the `switch` case, we send the current updated speed back to the computer via serial communication and finally set the motor speed.

The `Serial.read()` method will return 1 byte at a time while there is incoming data; this means that if you send from the Serial Monitor window a list of chars like ++++, it will increase the speed by four times.

It is like a computer-controlled motor with Arduino acting as a physical bridge.

Since the two projects we have seen in this chapter aren't really new, I would like to finish this chapter with a totally new example which, by the way, will introduce you to two new electronic components and a very powerful programming concept.

A computer connected dial thermometer

For this final example, I would like to build a device that displays the current temperature by using a dial and we will use two new components for it:

- A thermistor or temperature-dependent resistor as a sensor
- A servomotor as an output device to move the dial needle

Dealing with the thermistor is nothing new as it is just another kind of variable resistor and we will connect it to Arduino by once again creating a voltage divider. There is nothing new here.

Servomotors are a special kind of motors that don't freely spin, but are able to position at a specific angle and stay there instead thanks to a feedback mechanism and additional circuitry included in the motor case, and are depicted in the following image:

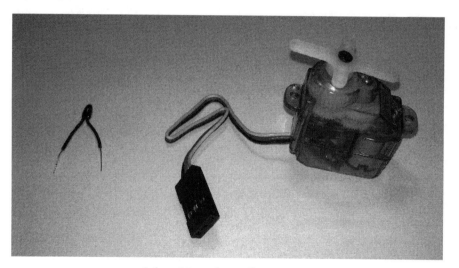

A thermistor and a small servomotor

The use of servomotors is very common in a multitude of projects but their management may be a bit tricky due to the fact that they need a specially forged train of pulses to operate.

Chapter 8

The Arduino language comes here to our rescue once again by incorporating a library to deal with servomotors, which makes connecting and programming one of these devices a breeze.

A library is simply a set of functions already elaborated and tested to accomplish a task. In this example, we will use the `servo` Arduino library to deal with the one we are going to connect.

The thermometer circuit

Here, you have the circuit we will use and its corresponding breadboard connections diagram.

As you see, we connect the thermistor just like the photocell or the potentiometer in our previous examples and this has been shown in the following image:

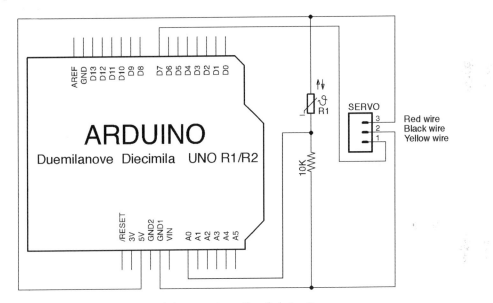

A thermometer with a dial circuit

Communicating with Others

The servomotor has three wires: V+, GND, and signal. We will connect the V+ wire (usually red) to **5V** and the GND (black) wire to **GND** from the Arduino power pin header, and finally the signal one (usually yellow or white) to pin **7**.

Perhaps you can see it better in the following breadboard connections diagram:

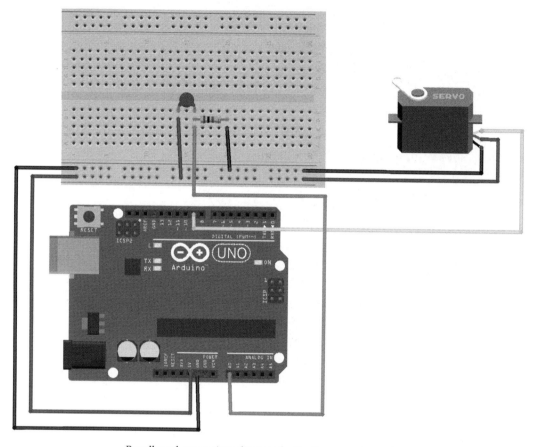

Breadboard connections diagram for the thermometer project

The code for the thermometer

Here, you have the complete code for the dial thermometer project. It shouldn't be hard to understand:

```
/*
 Chapter 08 - Communicating with others
 Dial thermometer
 By Francis Perea for Packt Publishing
*/

#include <Servo.h>

// Global Definitions
// The pin used for the transistor base
#define servoPin 7
// The increment for the motor speed
#define thermistor 0

// Global variables we will use
// A variable to store the read temperature
int temperature;
// and another to store the servo angle
int angle = 0;
// A  servo object
Servo aServo;

// Configuration of the board: three outputs and one input
void setup() {
  // Set the thermistor pin as an output
  pinMode(thermistor,INPUT); // optional
  // Init serial communication
  Serial.begin(9600);
  aServo.attach(7);
}

// Sketch execution loop
void loop(){
  // We read the value of the thermistor
  temperature = analogRead(thermistor);
```

```
    // Map it to a valid angle
    angle = map(temperature, 750, 850, 0, 179);
    // Send both back to the serial monitor
    Serial.print(temperature);
    Serial.print(" : ");
    Serial.println(angle);

    // Position the servomotor accordingly
    aServo.write(angle);

    // Wait some time to avoid the servo to vibrate
    delay(100);
}
```

The only new thing in this code is the programming of the servomotor.

We begin by including the library with the following:

```
#include <Servo.h>
```

This allows us to declare a new instance of a servo object in the `aServo` variable with this:

```
Servo aServo;
```

In the `setup()` function, we call the `aServo.attach(7)` method of our new object to specify through which pin the servomotor will be commanded.

Finally in the main loop, we simply map the received value from the thermistor into a valid angle in the range 0–179 and use the obtained value to position the servomotor with a call to `aServo.write(angle)`.

By sending the read value of the thermistor back to the computer via a serial communication, you can estimate the range of temperatures you want to control and allow the servo to represent. The proposed values 750–850 are the ones I found valid for my thermistor and the temperature ranges in my room, but I'm sure you will have to adapt to your specific configuration.

Here, you have an image of my particular assembly with a small gradient I printed to be used as the dial background:

A thermistor and a small servomotor

A very simple project but also a very visual one that I'm sure will make a perfect introduction to the use of servos and the servo library.

If you want to know more about servos and the Arduino servo library, you could visit the servomotor entry at the Wikipedia site at http://en.wikipedia.org/wiki/Servomotor and the Arduino **Reference** page for the servo library at http://arduino.cc/en/Reference/Servo.

Summary

Now you know how to make your Arduino talk serially to others, and use this commodity to help you correctly calibrate your sensors or let your projects receive data or even commands from your computer or other serial devices.

From the point of view of hardware, we have seen that almost every physical variable can be easily sensed into Arduino by using variable resistors in a voltage divider configuration.

We have also met the friendly servomotors and we have seen how easy it can be to incorporate one of these helpful motors in our projects.

And finally when talking about programming, we have seen what a library is and how much it can ease the work with different types of devices by providing ready-made functions, objects, and their methods at our disposal.

In the next chapter, we are going to meet interrupts that will allow our projects to immediately respond to external events, so let's not interrupt here and let's move on!

9
Dealing with Interrupts

In all our previous projects, we have been constantly looking for events to occur. We have been polling, but looking for events to occur involves a relatively big effort and a waste of CPU cycles to only notice that nothing happened.

In this chapter, we will learn about interrupts as a totally new way to deal with events, being notified about them instead of looking for them constantly.

Interrupts may be really helpful when developing projects in which fast or unknown events may occur, and thus we will see a very interesting project that will lead us to develop a digital tachograph for a computer-controlled motor.

Are you ready? Here we go!

The concept of an interruption

As you may have intuited, an interrupt is a special mechanism the CPU incorporates to have a direct channel to be noticed when some event occurs.

Most Arduino microcontrollers have two of these:

- Interrupt 0 on digital pin **2**
- Interrupt 1 on digital pin **3**

But some models, such as the Mega2560, come with up to five interrupt pins.

Once an interrupt has been notified, the CPU completely stops what it was doing and goes on to look at it, by running a special dedicated function in our code called **Interrupt Service Routine (ISR)**.

When I say that the CPU completely stops, I mean that even functions such as `delay()` or `millis()` won't be updated while the ISR is being executed.

Interrupts can be programmed to respond on different changes of the signal connected to the corresponding pin and thus the Arduino language has four predefined constants to represent each of these four modes:

- LOW: It will trigger the interrupt whenever the pin gets a LOW value
- CHANGE: The interrupt will be triggered when the pins change their values from HIGH to LOW or vice versa
- RISING: It will trigger the interrupt when the signal goes from LOW to HIGH
- FALLING: It is just the opposite of RISING; the interrupt will be triggered when the signal goes from HIGH to LOW

The ISR

The function that the CPU will call whenever an interrupt occurs is so important to the micro that it has to accomplish a pair of rules:

- They can't have any parameter
- They can't return anything
- The interrupts can be executed only one at a time

Regarding the first two points, they mean that we can neither pass nor receive any data from the ISR directly, but we have other means to achieve this communication with the function.

We will use global variables for it. We can set and read from a global variable inside an ISR, but even so, these variables have to be declared in a special way. We have to declare them as `volatile` as we will see later on in the code.

The third point, which specifies that only one ISR can be attended at a time, is what makes the function `millis()` not able to be updated. The `millis()` function relies on an interrupt to be updated, and this doesn't happen if another interrupt is already being served.

As you may understand, ISR is critical to the correct code execution in a microcontroller. As a rule of thumb, we will try to keep our ISRs as simple as possible and leave all heavyweight processing that occurs outside of it, in the main loop of our code.

The tachograph project

To understand and manage interrupts in our projects, I would like to offer you a very particular one, a tachograph, a device that is present in all our cars and whose mission is to account for revolutions, normally the engine revolutions, but also in brake systems such as an **Anti-lock Brake System (ABS)** and others.

Mechanical considerations

Well, calling it mechanical perhaps is too much, but let's consider how we are going to make our project account for revolutions.

For this example project, I have used a small DC motor driven through a small transistor and, like in lots of industrial applications, an encoded wheel is a perfect mechanism to read the number of revolutions. By simply attaching a small disc of cardboard perpendicularly to your motor shaft, it is very easy to achieve it.

By using our old friend, the optocoupler, we can sense something between its two parts, even with just a piece of cardboard with a small slot in just one side of its surface.

Here, you can see the template I elaborated for such a disc; the cross in the middle will help you position the disc as perfectly as possible, that is, the cross may be as close as possible to the motor shaft. The slot has to be cut off of the black rectangle as shown in the following image:

The template for the motor encoder

Once I printed it, I glued it to another piece of cardboard to make it more resistant and glued it all to the crown already attached to my motor shaft. If yours doesn't have a surface big enough to glue the encoder disc to its shaft, then perhaps you can find a solution by using just a small piece of dough or similar.

Once the encoder disc is fixed to the motor and spins attached to the motor shaft, we have to find a way to place the optocoupler in a way that makes it able to read through the encoder disc slot.

In my case, just a couple of drops of glue did the trick, but if your optocoupler or motor doesn't allow you to apply this solution, I'm sure that a pair of zip ties or a small piece of dough can give you another way to fix it to the motor too.

In the following image, you can see my final assembled motor with its encoder disc and optocoupler ready to be connected to the breadboard through alligator clips:

The complete assembly for the motor encoder

Once we have prepared our motor encoder, let's perform some tests to see it working and begin to write code to deal with interruptions.

A simple interrupt tester

Before going deep inside the whole code project, let's perform some tests to confirm that our encoder assembly is working fine and that we can correctly trigger an interrupt whenever the motor spins and the cardboard slot passes just through the optocoupler.

The only thing you have to connect to your Arduino at the moment is the optocoupler; we will now operate our motor by hand and in a later section, we will control its speed from the computer.

The test's circuit schematic is as follows:

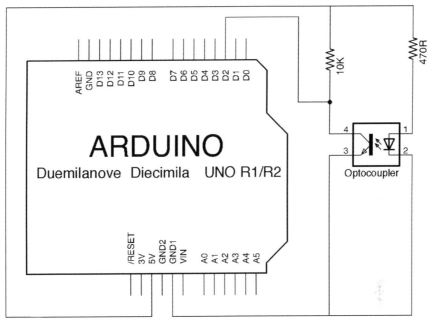

A simple circuit to test the encoder

Nothing new in this circuit, it is almost the same as the one used in the optical coin detector of *Chapter 5, Sensing the Real World through Digital Inputs*, with the only important and necessary difference of connecting the wire coming from the detector side of the optocoupler to pin **2** of our Arduino board, because, as said in the preceding text, the interrupt 0 is available only through that pin.

For this first test, we will make the encoder disc spin by hand, which allows us to clearly perceive when the interrupt triggers.

For the rest of this example, we will use the LED included with the Arduino board connected to pin **13** as a way to visually indicate that the interrupts have been triggered.

Our first interrupt and its ISR

Once we have connected the optocoupler to the Arduino and prepared things to trigger some interrupts, let's see the code that we will use to test our assembly.

The objective of this simple sketch is to commute the status of an LED every time an interrupt occurs. In the proposed tester circuit, the LED status variable will be changed every time the slot passes through the optocoupler:

```
/*
 Chapter 09 - Dealing with interrupts
 A simple tester
 By Francis Perea for Packt Publishing
*/

// A LED will be used to notify the change
#define ledPin 13

// Global variables we will use
// A variable to be used inside ISR
volatile int status = LOW;

// A function to be called when the interrupt occurs
void revolution(){
  // Invert LED status
  status=!status;
}

// Configuration of the board: just one output
void setup() {
  pinMode(ledPin, OUTPUT);
  // Assign the revolution() function as an ISR of interrupt 0
  // Interrupt will be triggered when the signal goes from
  // LOW to HIGH
  attachInterrupt(0, revolution, RISING);
}

// Sketch execution loop
void loop(){
  // Set LED status
  digitalWrite(ledPin, status);
}
```

Let's take a look at its most important aspects.

The LED pin apart, we declare a variable to account for changes occurring. It will be updated in the ISR of our interrupt; so, as I told you earlier, we declare it as follows:

```
volatile int status = LOW;
```

Following which we declare the ISR function, revolution(), which as we already know doesn't receive any parameter nor return any value. And as we said earlier, it must be as simple as possible. In our test case, the ISR simply inverts the value of the global volatile variable to its opposite value, that is, from LOW to HIGH and from HIGH to LOW.

To allow our ISR to be called whenever an interrupt 0 occurs, in the setup() function, we make a call to the attachInterrupt() function by passing three parameters to it:

- **Interrupt**: The interrupt number to assign the ISR to
- **ISR**: The name without the parentheses of the function that will act as the ISR for this interrupt
- **Mode**: One of the following already-explained modes that defines when exactly the interrupt will be triggered

In our case, the concrete sentence is as follows:

```
attachInterrupt(0, revolution, RISING);
```

This makes the function revolution() be the ISR of interrupt 0 that will be triggered when the signal goes from LOW to HIGH.

Finally, in our main loop there is little to do. Simply update the LED based on the current value of the status variable that is going to be updated inside the ISR.

If everything went right, you should see the LED commute every time the slot passes through the optocoupler as a consequence of the interrupt being triggered and the revolution() function inverting the value of the status variable that is used in the main loop to set the LED accordingly.

A dial tachograph

For a more complete example in this section, we will build a tachograph, a device that will present the current revolutions per minute of the motor in a visual manner by using a dial.

The motor speed will be commanded serially from our computer by reusing some of the codes in our previous projects.

It is not going to be very complicated if we include some way to inform about an excessive number of revolutions and even cut the engine in an extreme case to protect it, is it?

The complete schematic of such a big circuit is shown in the following image. Don't get scared about the number of components as we have already seen them all in action before:

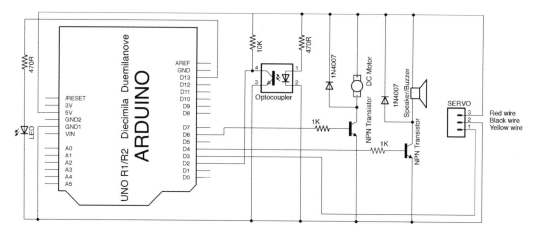

The tachograph circuit

As you may see, we will use a total of five pins of our Arduino board to sense and command such a set of peripherals:

- **Pin 2**: This is the interrupt 0 pin and thus it will be used to connect the output of the optocoupler.
- **Pin 3**: It will be used to deal with the servo to move the dial.
- **Pin 4**: We will use this pin to activate the sound alarm once the engine current has been cut off to prevent overcharge.
- **Pin 6**: This pin will be used to deal with the motor transistor that allows us to vary the motor speed based on the commands we receive serially. Remember to use a PWM pin if you choose to use another one.

- **Pin 13**: Used to indicate with an LED an excessive number of revolutions per minute prior to cutting the engine off.

There are also two more pins that, although not physically connected, will be used, pins 0 and 1, given that we are going to talk to the device serially from the computer.

Breadboard connections diagram

There are some wires crossed in the previous schematic, and perhaps you can see the connections better in the following breadboard connection image:

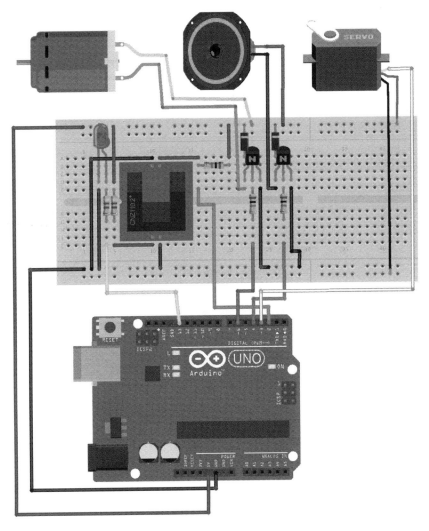

Breadboard connection diagram for the tachograph

The complete tachograph code

This is going to be a project full of features and that is why it has such a number of devices to interact with. Let's resume the functioning features of the dial tachograph:

- The motor speed is commanded from the computer via a serial communication with up to five commands:
    - Increase motor speed (+)
    - Decrease motor speed (-)
    - Totally stop the motor (0)
    - Put the motor at full throttle (*)
    - Reset the motor after a stall (R)
- Motor revolutions will be detected and accounted by using an encoder and an optocoupler
- Current revolutions per minute will be visually presented with a dial operated with a servomotor
- It gives a visual indication via an LED of a high number of revolutions
- In case a maximum number of revolutions is reached, the motor current will be cut off and an acoustic alarm will sound

With such a number of features, it is normal that the code for this project is going to be a bit longer than our previous sketches. Here is the code:

```
/*
 Chapter 09 - Dealing with interrupt
 Complete tachograph system
 By Francis Perea for Packt Publishing
*/

#include <Servo.h>

//The pins that will be used
#define ledPin 13
#define motorPin 6
#define buzzerPin 4
#define servoPin 3

#define NOTE_A4 440
// Milliseconds between every sample
#define sampleTime 500
```

```
// Motor speed increment
#define motorIncrement 10
// Range of valir RPMs, alarm and stop
#define minRPM   0
#define maxRPM 10000
#define alarmRPM 8000
#define stopRPM 9000

// Global variables we will use
// A variable to be used inside ISR
volatile unsigned long revolutions = 0;
// Total number of revolutions in every sample
long lastSampleRevolutions = 0;
// A variable to convert revolutions per sample to RPM
int rpm = 0;
// LED Status
int ledStatus = LOW;
// An instace on the Servo class
Servo myServo;
// A flag to know if the motor has been stalled
boolean motorStalled = false;
// The current dial angle
int dialAngle = 0;
// A variable to store serial data
int dataReceived;
// The current motor speed
int speed = 0;
// A time variable to compare in every sample
unsigned long lastCheckTime;

// A function to be called when the interrupt occurs
void revolution(){
  // Increment the total number of
  // revolutions in the current sample
  revolutions++;
}

// Configuration of the board
void setup() {
  // Set output pins
  pinMode(motorPin, OUTPUT);
```

```
  pinMode(ledPin, OUTPUT);
  pinMode(buzzerPin, OUTPUT);
  // Set revolution() as ISR of interrupt 0
  attachInterrupt(0, revolution, CHANGE);
  // Init serial communication
  Serial.begin(9600);
  // Initialize the servo
  myServo.attach(servoPin);
  //Set the dial
  myServo.write(dialAngle);
  // Initialize the counter for sample time
  lastCheckTime = millis();
}

// Sketch execution loop
void loop(){
  // If we have received serial data
  if (Serial.available()) {
    // read the next char
    dataReceived = Serial.read();
    // Act depending on it
    switch (dataReceived){
      // Increment speed
      case '+':
        if (speed<250) {
          speed += motorIncrement;
        }
        break;
      // Decrement speed
      case '-':
        if (speed>5) {
          speed -= motorIncrement;
        }
        break;
      // Stop motor
      case '0':
        speed = 0;
        break;
      // Full throttle
      case '*':
        speed = 255;
        break;
```

```
        // Reactivate motor after stall
      case 'R':
        speed = 0;
        motorStalled = false;
        break;
    }
    //Only if motor is active set new motor speed
    if (motorStalled == false){
      // Set the motor speed
      analogWrite(motorPin, speed);
    }
  }
  // If a sample time has passed
  // We have to take another sample
  if (millis() - lastCheckTime > sampleTime){
    // Store current revolutions
    lastSampleRevolutions = revolutions;
    // Reset the global variable
    // So the ISR can begin to count again
    revolutions = 0;
    // Calculate revolution per minute
    rpm = lastSampleRevolutions * (1000 / sampleTime) * 60;
    // Update last sample time
    lastCheckTime = millis();
    // Set the dial according new reading
    dialAngle = map(rpm,minRPM,maxRPM,180,0);
    myServo.write(dialAngle);
  }
  // If the motor is running in the red zone
  if (rpm > alarmRPM){
    // Turn on LED
    digitalWrite(ledPin, HIGH);
  }
  else{
    // Otherwise turn it off
    digitalWrite(ledPin, LOW);
  }
  // If the motor has exceeded maximum RPM
  if (rpm > stopRPM){
    // Stop the motor
    speed = 0;
    analogWrite(motorPin, speed);
```

```
      // Disable it until an 'R' command is received
      motorStalled = true;
      // Make alarm sound
      tone(buzzerPin, NOTE_A4, 1000);
    }
    // Send data back to the computer
    Serial.print("RPM: ");
    Serial.print(rpm);
    Serial.print(" SPEED: ");
    Serial.print(speed);
    Serial.print(" STALL: ");
    Serial.println(motorStalled);
  }
```

It is the first time in this book that I think I have nothing to explain regarding the code that hasn't been already explained before.

I have commented everything so that the code can be easily read and understood.

In general lines, the code declares both constants and global variables that will be used and the ISR for the interrupt.

In the setup section, all initializations of different subsystems that need to be set up before use are made: pins, interrupts, serials, and servos.

The main loop begins by looking for serial commands and basically updates the speed value and the stall flag if command R is received.

The final motor speed setting only occurs in case the stall flag is not on, which will occur in case the motor reaches the `stopRPM` value.

Following with the main loop, the code looks if it has passed a sample time, in which case the revolutions are stored to compute real **revolutions per minute** (**rpm**), and the global revolutions counter incremented inside the ISR is set to 0 to begin again.

The current rpm value is mapped to an angle to be presented by the dial and thus the servo is set accordingly.

Next, a pair of controls is made:

- One to see if the motor is getting into the red zone by exceeding the max `alarmRPM` value and thus turning the alarm LED on
- And another to check if the `stopRPM` value has been reached, in which case the motor will be automatically cut off, the `motorStalled` flag is set to `true`, and the acoustic alarm is triggered

When the motor has been stalled, it won't accept changes in its speed until it has been reset by issuing an `R` command via serial communication.

In the last action, the code sends back some info to the Serial Monitor as another way of feedback with the operator at the computer and this should look something like the following screenshot:

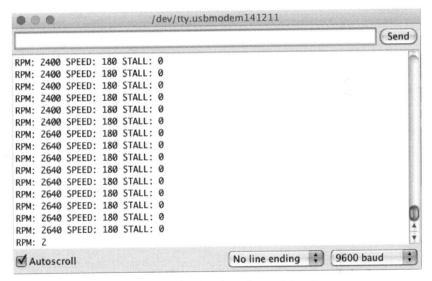

Serial Monitor showing the tachograph in action

Modular development

It has been quite a complex project in that it incorporates up to six different subsystems: optocoupler, motor, LED, buzzer, servo, and serial, but it has also helped us to understand that projects need to be developed by using a modular approach.

We have worked and tested every one of these subsystems before, and that is the way it should usually be done.

By developing your projects in such a submodular way, it will be easy to assemble and program the whole of the system.

As you may see in the following screenshot, only by using such a modular way of working will you be able to connect and understand such a mess of wires:

A working desktop may get a bit messy

Summary

I'm sure you have got the point regarding interrupts with all the things we have seen in this chapter.

We have met and understood what an interrupt is and how the CPU attends to it by running an ISR, and we have even learned about their special characteristics and restrictions and that we should keep them as few as possible.

On the programming side, the only thing necessary to work with interrupts is to correctly attach the ISR with a call to the `attachInterrupt()` function.

From the point of view of hardware, we have assembled an encoder that has been attached to a spinning motor to account for its revolutions.

Finally, we have the code. We have seen a relatively long sketch, which is a sign that we are beginning to master the platform, are able to deal with a bigger number of peripherals, and that our projects require more complex software every time we have to deal with these peripherals and to accomplish all the other necessary tasks to meet what is specified in the project specifications.

In the next and final chapter, we will see another real-case application for Arduino, controlling a greenhouse this time.

10
Arduino in a Real Case – Greenhouse Control

At this point in the book, we have already seen every important aspect of the Arduino platform, from its inputs and outputs to the use of interrupts and communications.

For this final chapter, I would like to propose a complete project that uses as many concepts as possible that have been seen until now, and I thought that a greenhouse controller could be a good example.

We will even meet a final component, the relay, which will help us deal with external devices not directly connected to our circuitry.

So, let's go into our final project and see how much Arduino can do by simply connecting a bunch of electronic components and writing a little sketch.

A greenhouse controller

This project aims to control and automate all aspects in a theoretical greenhouse that has a number of sensors, which provide information regarding the environment and act on a number of devices as a reaction to changes in its environment.

To begin designing, connecting, and programming our controller, let's begin by clearly setting out what we expect of such a system by creating a simple list of the functions it should accomplish.

The controller requirements

In general terms, the controller should monitor and control the most common variables that could affect such an installation:

- Air temperature
- Soil humidity
- Direct solar lighting

Secondly, the controller will also offer an alarm mechanism that could be manually triggered by a person inside the greenhouse in case of an emergency.

Modular design

To accomplish the previous requirements, we will divide the complete greenhouse controller into four main different modules or subsystems:

- **Temperature**: It is perhaps the most important subsystem. Its main objective is to constantly monitor the temperature inside the greenhouse and operate a different number of devices to maintain the temperature between the limits of a pre-established secure range.
- **Watering**: It will be responsible for keeping the humidity of the soil between the preset limits by operating an external water pump.
- **Lighting**: The controller will open a retractable roof during the day and close it at night by using a big servomotor that will rotate an arm that moves the roof panels.
- **Alarm**: In case a greenhouse operator needs it, he/she could press a button to activate a visual and acoustics alarm to indicate an emergency situation to others.

Let's take a deeper look at every one of these subsystems to try to understand their functioning and the electronic components we will use to accomplish these tasks.

Temperature control

The temperature control subsystem will receive information through a simple thermistor and acts on up to four different devices to control the temperature:

- **Visual indicator**: The temperature should be constantly announced via a visual panel. In our case, the panel is going to be replaced by three simple LEDs that will act as a visual level of the temperature inside the greenhouse.
- **Fan**: As the temperature goes up, the controller will act on a fan's speed, a motor in our example, to allow the greenhouse to be ventilated.
- **Retractable roof**: The greenhouse roof can be opened or closed, and thus the temperature control subsystem can open it in case the temperature is getting too hot.
- **Watering pump**: In case of an excessive temperature, the control system can activate the watering to increase the humidity of the greenhouse or try to avoid a fire inside the greenhouse.

Humidity control

The monitoring of the soil's humidity can be done thanks to special humidity sensors that can be bought in any major gardening store (around $2) or built by simply using a pair of nails as you can see, for example, in `http://www.instructables.com/id/Garduino-Gardening-Arduino/step4/Build-Your-Moisture-Sensor/`, but for the sake of simplicity in the project schematic and code, I will replace it by a simple potentiometer that will represent any generic resistive sensor.

We will also consider that the greenhouse counts on an external watering system that has its own pump. From the point of view of our controller, we only have to switch it on, and for this purpose we will use a new electronic component, a relay, that will be introduced in the *The relay as a mediator* section.

Lighting control

Thanks to the retractable roof, the greenhouse allows you to expose the plants to direct solar light during the day. In the night hours, the roof will be closed to avoid freezing.

From the point of view of the controller, we will only have to position the servo at 0 degrees to close the roof or at 180 degrees to open it.

Manual alarm

By using a simple push button, the greenhouse operator can activate the system alarm in case of an emergency.

The alarm routine will trigger up to four actions:

- A loud sound will be produced to give an acoustic indication of the alarm situation
- Panel LEDs will blink instead of indicating the temperature level
- The retractable roof will be closed in case it is open to prevent air from flowing in case a fire breaks out
- The watering system will be activated to try to increase the humidity in the greenhouse

Input and output devices

Once we know what is expected of such a system, let's make an account of the input and output devices that will be needed to accomplish the aforementioned requirements.

From the point of view of inputs, we will need to connect the following components to act as sensors:

- **Thermistor**: To allow for temperature monitoring in a similar way as we have already seen in *Chapter 8, Communicating with Others*.
- **Photocell**: To sense direct sunlight over the greenhouse and allow opening or closing of the retractable roof.
- **Humidity resistive sensor**: To sense the soil's humidity and trigger the watering pump. As mentioned before, this sensor will be replaced in the schematics and code by a simple potentiometer to allow for simple testing.
- **Push button**: This will allow the greenhouse operator to trigger the general alarm.

On the other side, talking about outputs, our project will use the following components connected to Arduino outputs:

- **Motor**: To operate the fan that will ventilate the greenhouse
- **Servomotor**: To open or close the retractable roof
- **Relay**: To activate the watering pump, as we will see in the next section
- **Buzzer**: To produce the alarm sound

Chapter 10

The relay as a mediator

A relay is an electromechanical device also known as an electric switch that, thanks to the magnetic field generated by any electrical current, allows opening or closing a mechanical switch placed very close to a small coil just by powering the coil.

In a typical relay, the coil and the switch are encapsulated in a small case, offering usually four terminals, two for powering the coil and two for the switch.

Some types of relays even offer three terminals for the switch side as follows:

- A normally closed terminal
- A normally open terminal
- A common terminal

In the next figure, you can see the schematic symbol of a relay and an image of the relay I've used for assembling the current project:

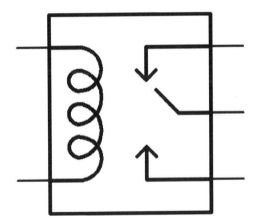 

A relay symbol and a picture of a real relay

The most common use of relays is to operate external or different voltage circuits without connecting and powering them to the controller circuit. In our case, we will use it to activate an alternate current watering pump by simply connecting the switch part of the relay as a usual manual switch inserted in one of the power wires of the watering pump.

From the point of view of managing a relay from Arduino, a coil is just a new inductive load, like that on a motor or a speaker, so we will connect it just the same way as we did with these other components as you can see in the following schematic:

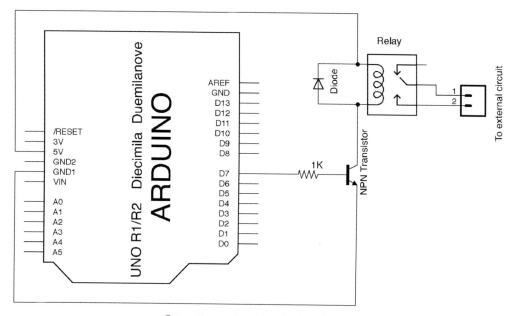

Connecting a relay through a transistor

I have also used a diode to protect against the possible counter-electromotive force peaks that could be produced as we did with the motor and the speaker.

For the sake of simplicity in the schematic and breadboard connections diagram, I will use only two-wire terminal connectors in the switch part of the relay so that the two wires coming from the watering pump can be easily connected to the relay.

Regarding its programming, we can power the coil on and off by simply setting the pin to which the transistor driver base is connected to HIGH or LOW. It doesn't have to be a PWM pin; any digital pin can do the work, because we are not going to use different voltage levels, just on or off.

For more detailed information relating to relays, its functioning, use, and different types, you could go and visit the Wikipedia entry for the relay at http://en.wikipedia.org/wiki/Relay.

Chapter 10

The greenhouse controller circuit

Once we have met the relay as the new component that we will use in this project, here you have the complete schematic of the project.

Once again, don't let the number of components scare you. If you study the circuit one component at a time, you will immediately notice that they are all already known and their connections have been explained and tested before. This has been shown in the following image:

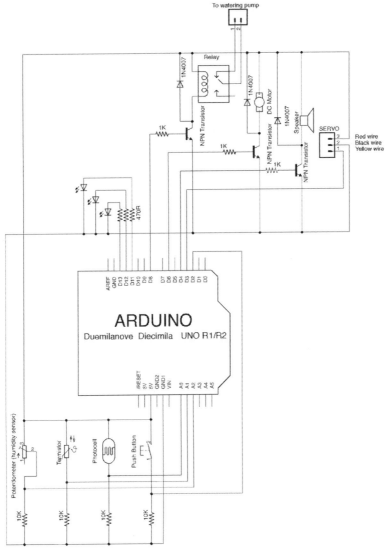

Complete greenhouse controller circuit schematic

[173]

At this point and before entering the code explanation of this project, I would like to make a compilation of the pins we will be using for the project, in the assembly of the circuit and in the code, to have a clear understanding about which kind of pin is necessary for every one of the connected components.

Regarding inputs, we will have the photocell, thermistor, and potentiometer connected to analog inputs because they are going to be read as analog values.

The other input device, the push button, will be connected to digital pin **2** because we will read it by using Arduino interrupt 0, which is only available through this pin. This way, we give total preference to the alarm by stopping anything else that the processor would be doing when the button is pressed, and we ensure that the button pressing is recorded even if Arduino is doing anything else.

Talking about outputs, the only device that requires an analog output and thus will be connected to a PWM pin is the motor, because we will be varying its speed in an analog way.

The other devices (relay, speaker, servomotor, and LEDs) can be connected to any digital pin, be it PWM or not, because we will be dealing with them as simple digital outputs and only generating values of HIGH or LOW with the `digitalWrite()` function.

To simply summarize the pins used for this project, here you have the full connections list:

- **Analog 0**: Photocell
- **Analog 1**: Thermistor
- **Analog 2**: Humidity sensor, the potentiometer in our schematic
- **Digital 2**: Button to be read through interrupt 0
- **Digital 3**: Servomotor
- **Digital 4**: Speaker
- **Digital 6**: Motor to be used with PWM
- **Digital 8**: Relay
- **Digital 11**: Green LED
- **Digital 12**: Yellow LED
- **Digital 13**: Red LED

It may seem a big number of used pins, but let me say that we still have nine available pins:

- Three more free inputs in the analog input side
- Six digital pins, three of which are PWM

You could still include some more subsystems to the controller like, for example, a fire extinguisher operated through a relay just like the watering pump, by using just another digital pin.

Who said Arduino had few available pins? And, of course, you can still use an Arduino Mega if your projects need even more pins.

Breadboard connections diagram

Here, you have the breadboard connections diagram for the complete greenhouse controller project:

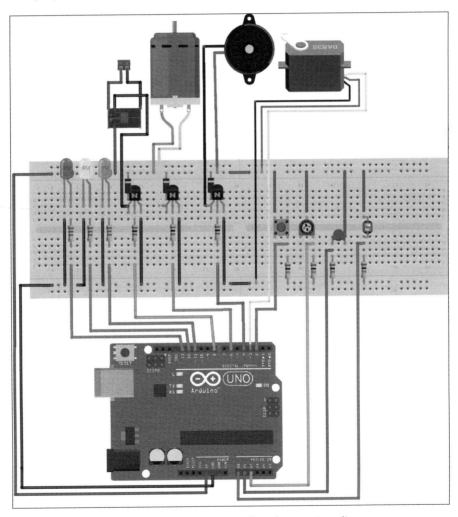

Greenhouse controller project breadboard connections diagram

For this project diagram, I had to use a full-sized breadboard, and even so there are some components that aren't correctly seen because they are beside others. In any case, I'm sure that at this point of the book you don't really need this diagram and you will be able to assemble your circuit directly from the schematic.

Please notice the two wires connecting the two halves of the power rails just in the middle. In a full-sized breadboard, the upper and lower rails are not connected all along. They are divided into two halves to allow for different power sources. In our example, we don't need this and will power all the different devices from the same power source, so you have to connect both halves of the rails by simply using a small wire or a jumper; otherwise, all the components on the other half won't receive any power.

The greenhouse controller code

Since the code of this project is perhaps a little longer than the previous examples, I will present it first by sections without comments and include the complete code for reference at the end of this chapter.

Libraries and constant definitions

We begin our code by including the `servo` library that we will use to deal with the servomotor that will operate the retractable roof. We define constants that will be used all along the code as follows:

```
#include <Servo.h>

#define redLedPin 13
#define yellowLedPin 12
#define greenLedPin 11
#define relayPin 8
#define motorPin 6
#define buzzerPin 4
#define servoPin 3
#define buttonPin 2
#define potentiometerPin 2
#define thermistorPin 1
#define photocellPin 0
#define NOTE_A4 440
#define NOTE_A3 220
#define TEMPMIN 820
```

```
#define TEMPMAX 850
#define MEDIUMPOWER 128
#define FULLPOWER 255
#define BEGINWATERING 700
#define ENDWATERING 550
#define DAYLIGHT 400
```

Here is a breakdown of the 20 constants used in the preceding code:

- The first 11 constants are simply to reference the pins that we will use to connect the different devices. They are all lowercase with every word's first letter in uppercase and they all finish in `Pin`.
- From there on, there are some constants that will be used mainly to compare their values in conditional statements all along the code.
- `NOTE_A3` and `NOTE_A4` define the frequency of the notes that will be produced in the acoustic alarm.
- `TEMPMAX` and `TEMPMIN` represent the range of temperatures in which the controller can operate. You will have to adapt them to the readings your thermistor gives.
- `MEDIUMPOWER` and `FULLPOWER` are the values at which the motor will spin in the medium and high temperatures.
- `BEGINWATERING` and `ENDWATERING` are the values between which the watering will occur. We will activate the water pump when the soil humidity sensor or the potentiometer gives a value of `700` and end the watering when the humidity raises and makes the sensor lower its internal resistance to give a value of `550`.
- Finally, `DAYLIGHT` is the photocell value that indicates that the night is off and that we can open the retractable roof.

Global variables

There aren't really too many variables in this example, just those needed to read the analog sensors and three more:

```
volatile boolean buttonPressed = false;
int tempValue;
int humidityValue;
int lightValue;
Servo myServo;
int state;
```

Here is a breakdown of the variables used in the preceding code:

- The volatile Boolean `buttonPressed` is, as you may have guessed, the one we will be using inside the **Interrupt Service Routine (ISR)** for the button press. It is simply a Boolean that we will use to flag up the button pressing.
- `tempValue`, `humidityValue`, and `lightValue` are used, as I said before, to store the values read from every one of the analog sensors.
- The `myServo` variable is an instance of the `Servo` class that will allow us to operate the servomotor.
- Finally, `state` is a variable used in a `switch` control structure to activate each subsystem depending on its value. We will see it later in action when we study the main loop code.

The interrupt ISR

As mentioned before, the button pressing will be managed via an interrupt and thus an ISR has to be written. We saw in *Chapter 9, Dealing with Interrupts*, that the code inside the ISR should be as concise as possible avoiding big processing inside this critical function.

In our case, we simply set the `buttonPressed` Boolean variable to `true`, which will execute some code inside the main loop once this is detected.

The complete code of this function is as follows:

```
void buttonPress(){
  buttonPressed = true;
}
```

The alarm routine

We talked before about all the things that should be unchained once an alarm situation is detected.

Since this alarm situation may occur in more than one place in the code, I have decided to include all these actions in a function and call it whenever it is needed.

The `alarmRoutine()` function code is as follows:

```
void alarmRoutine(){
  myServo.write(0);
  digitalWrite(relayPin,HIGH);
  tone(buzzerPin,NOTE_A4);
  digitalWrite(greenLedPin,HIGH);
  digitalWrite(yellowLedPin,HIGH);
  digitalWrite(redLedPin,HIGH);
  delay(1000);
  tone(buzzerPin,NOTE_A3);
  digitalWrite(greenLedPin,LOW);
  digitalWrite(yellowLedPin,LOW);
  digitalWrite(redLedPin,LOW);
  delay(1000);
  noTone(buzzerPin);
}
```

As you may have understood at this point, we simply execute sequentially all the actions of the alarm routine, that is:

1. Close the roof by setting the servo at 0 degrees.
2. Activate the watering system by setting a HIGH value at the relay pin, or to be precise at the pin that connects to the base of the transistor that acts as a driver to the relay.
3. Produce a tone.
4. Turn all the three LEDs on.
5. Wait for a second.
6. Produce a different tone in a similar way to a real alarm.
7. Turn all the three LEDs off.
8. Wait for another second.
9. Turn the sound off.

Arduino in a Real Case – Greenhouse Control

Initialization and board configuration

The next step in our code is to configure the used pins and other necessary setup routines for all used devices. Here is the complete `setup()` function of our project sketch:

```
void setup() {
  pinMode(redLedPin, OUTPUT);
  pinMode(yellowLedPin, OUTPUT);
  pinMode(greenLedPin, OUTPUT);
  pinMode(relayPin,OUTPUT);
  pinMode(motorPin, OUTPUT);
  pinMode(buzzerPin, OUTPUT);
  myServo.attach(servoPin);
  attachInterrupt(0, buttonPress, RISING);
  Serial.begin(9600);
}
```

The `pinMode()` functions call apart, we simply initialize our servo instance, set the ISR for the interrupt 0, and begin a serial communication that will help us calibrate our sensors.

There is nothing new here.

The main execution loop

Finally, here is the main loop of our sketch where it all happens. You will immediately notice that I simply check every subsystem in sequence and trigger every action needed, so let's analyze every one of the four subsystems in the code.

Temperature subsystem

The code corresponding to the temperature control is as follows:

```
tempValue = analogRead(thermistorPin);
  state =map(tempValue, TEMPMIN, TEMPMAX, 1, 4);
  switch (state){
    case 1:
     digitalWrite(greenLedPin,HIGH);
     break;
    case 2:
      digitalWrite(yellowLedPin,HIGH);
      analogWrite(motorPin, MEDIUMPOWER);
      break;
    case 3:
```

```
          digitalWrite(redLedPin, HIGH);
          analogWrite(motorPin,FULLPOWER);
          myServo.write(180);
          break;
        case 4:
          alarmRoutine();
          break;
      }
```

We begin by reading the sensor and storing its value in the `tempValue` variable.

From now on, we map the read value into the `state` variable. The mapping will convert the read value, valid between the `TEMPMIN` and `TEMPMAX` constants that we presented earlier, into a new range of 1 to 4 that will ease the process of selecting what to do depending on one of the four possible states.

We use this variable to select a case up of four in the `switch` control structure that will represent the four possible states of our system and which will trigger different actions:

- **State 1**: It is all under control, and we simply set the green LED on.
- **State 2**: The temperature is rising, so we turn yellow LED on and the motor at `MEDIUMPOWER`.
- **State 3**: The temperature is getting too hot, so we indicate it by turning the red LED on, power the motor at `FULLPOWER`, and open the retractable roof to try to lower the temperature by setting the servo at 180 degrees.
- **State 4**: This state represents an emergency, so, we simply call the already presented `alarmRoutine()` function.

Humidity subsystem

After the temperature control has been made, we are going to see how the humidity control is made. This case is quite simple because it only has two possible situations: if the reading is above the `BEGINWATERING` value, we open the watering pump and we close it when the reading is below the `ENDWATERING` value:

```
     humidityValue=analogRead(potentiometerPin);
       if (humidityValue>BEGINWATERING){
         digitalWrite(relayPin,HIGH);
       }
       if (humidityValue<ENDWATERING){
         digitalWrite(relayPin,LOW);
       }
```

Notice how these two conditionals are independent, meaning that the one that activates the watering will be true first and quite a good number of loops later the other will be true.

This is so because most resistive humidity sensors offer a lower resistance when the humidity increases. This way, when the soil is really dry, we will trigger the water pump and will leave it on for as much time as needed until a new reading of the sensor confirms that the humidity has increased and so the sensor reading has decreased.

Lighting subsystem

The photocell control is very similar to the humidity one, but in this case both possible cases are exclusive, which means that either it is day or it is night; there is no range here, it is only a simple cutoff value we all have, and thus we either open the roof or we close it.

Here, there is only an `if` conditional statement with two exclusive branches, a totally different approach to the humidity control:

```
lightValue = analogRead(photocellPin);
  if (lightValue>DAYLIGHT){
    myServo.write(0);
  }
  else{
    myServo.write(180);
  }
```

Just as in the previous section, we read the sensor and compare it with the `DAYLIGHT` cutoff value to open or close the roof by setting the servomotor at `0` or `180` degrees.

Alarm subsystem

The alarm control is quite simple; it just checks the `buttonPressed` variable to see if it was changed in the ISR, in which case it resets it and calls the `alarmRoutine()` function:

```
if (buttonPressed){
   buttonPressed = false;
   alarmRoutine();
 }
```

Serial feedback and calibration

To allow for calibrating the sensor, we finish the main loop by sending back the reading of every analog sensor and the mapped state to the Serial Monitor.

Once the project runs, this code can be simply commented out to avoid working on unneeded instructions:

```
Serial.print(" S: ");
Serial.print(state);
Serial.print(" T: ");
Serial.print(tempValue);
Serial.print(" H: ");
Serial.print(humidityValue);
Serial.print(" L: ");
Serial.println(lightValue);
```

The complete project code

You can download the complete code for this project from the Packt Publishing website. The name of the file is `_17_greenhouse.ino` inside the `8569_10_Code` folder.

Final considerations

This was a relatively simple project based on a theoretical situation, but if you are really interested in working on a real gardening project, you should take a look on the Internet because there are plenty of similar projects and ideas that might help you.

In particular, I liked the initiative of the GardenBot project at `http://gardenbot.org`, an open source initiative to build a complete monitoring and automation gardening system.

I'm sure you will like to take a look at it; you will enjoy seeing how much you can understand of such a project after reading this book and it could even give you some ideas to improve the project in this chapter.

Summary

With this chapter, we finish our journey through the Arduino platform, but even in this last chapter, we have met some new components such as the relay.

Once again, we have seen the importance of a modular design when developing these kinds of projects and how much the code benefits from such an approach.

I'm sure that with all that you have learned in this book, you will be ready to develop your own Arduino-based projects and that you will have developed your own ideas for projects while you read this book.

So, don't put this book too far away from your working desktop, so that you can use it as a reference and begin to enjoy working on all these projects by yourself.

Index

A

accounting
  micros() function 112
  millis() function 112
  versus stopping 112
Adafruit
  URL 23
alarm, greenhouse controller
  manual alarm 170
ambient light meter
  about 97-100
  breadboard connections 100
  circuit 134-138
  code 101
analog
  analogWrite() function 64
  connections diagram 63
  controlling, through code 64
  fading LED code, completing 66-68
  for loop 64, 65
  output circuit 62, 63
  signals 61, 62
analog sensors 78
analog signals
  versus digital signals 43, 44
Analog to Digital Converter (ADC) 95
analog values
  sensing 95
analogWrite() function 64
Anti-lock Brake System (ABS) 151
Arduino
  about 7, 8
  data, sending 139-141
  development environment, running 32, 33
  download section, URL 28
  drivers, installing 30
  Esplora 19
  family members 19
  features 8, 9
  features, URL 9
  Getting Started guide, URL 30
  Integrated Development Environment (IDE) 8
  language 42
  LilyPad Arduino 14-16
  micro board 18, 19
  mini board 18, 19
  nano board 18, 19
  package, downloading 28
  playground section, URL 24
  playground tab, URL 32
  rules 24
  shields 23
  software, installing 29
  store, URL 11
  time control, functions 111
  traditional forum, URL 24
  unofficial boards 22, 23
  users, teaching users 24
  variable resistor, connecting to 98
Arduino Due 22
Arduino Ethernet
  about 12, 13
  URL 14
Arduino Mega 2560
  about 11
  URL 12
Arduino Robot
  about 20, 21
  product page, URL 21

Arduino Uno
  about 9, 10
  URL 11
Arduino Yún
  about 16-18
  URL 18
attachInterrupt() function 155
  Interrupt 155
  ISR 155
  Mode 155

B

beats per minute (bpm)
  URL 128
breadboard connections, diagram
  for ambient light meter circuit 100
  for DC motor speed control 107
  for digital input 83
  for LED circuit 48, 49
  for optical switch 90, 91
  using 45, 46
button
  connecting, as digital input 78, 79
  pressed 79
  released 79

C

circuit
  about 44
  asymmetric blinking code 49-51
  breadboard connections, diagram 48, 49
  breadboard, using 45
  C language, syntax considerations 52
  faults, troubleshooting 52, 53
  LED circuit 46, 47
  limit per pin 55-58
  multiple outputs, dealing with 53, 54
  schematic 47
circuit schematic 82
C language
  syntax, considerations 52
code
  writing, for press reaction 84
code editor 34, 35
coin detector
  optocoupler, using as 89

  schematic 89, 90
conditional bifurcation
  about 84
  decisions, making with 86
Cooking Hacks
  URL 24

D

DAC
  code 95
data
  sending, to Arduino 138-141
DC motor speed control
  about 105
  breadboard connections diagram 107
  code 108
  potentiometer 106
  schematic 106
debouncing
  about 87
  URL 87
development environment, Arduino
  code editor 34, 35
  message area 36
  running 32, 33
  toolbar 34
dial tachograph
  about 156, 157
  breadboard connections, diagram 157
  code 158-163
  modular development 164
dial thermometer
  circuit 143, 144
  code 145-147
  computer connected 142, 143
digital input
  breadboard connections, diagram 83
  button, connecting as 78, 79
  configuring 85
  decisions, making with conditional
      bifurcations 86
  digitalRead(button) 85
  example code 92
  pinMode(button,INPUT) 85
  reading 85

digital sensors
  about 78
  types 87, 88
digital signals
  versus analog signals 43, 44
Digital to Analog Converter (DAC)
  about 61
  URL 62
drivers, installing
  about 30
  for Linux 32
  for Mac OS X 32
  for Windows 31, 32

E

EAGLE software
  URL 75
Electromagnetic Force (EMF) 70
Esplora
  about 19
  URL 20
Examples command 40

F

for loop
  about 64, 65
  Condition 65
  Increment 65
  Initialization 65
  URL 65
Fritzing
  URL 48
FTDI
  manufacturer page, URL 31
  Uno driver, URL 32

G

GardenBot project
  URL 183
GND 144
greenhouse controller
  about 167
  breadboard connections, diagram 175, 176
  circuit 173-175
  input and output devices 170
  modular design 168
  requisites 168
  temperature control 169
greenhouse controller, code
  about 176
  alarm control 182
  alarm routine 178, 179
  considerations 183
  execution loop 180
  global variables 177, 178
  humidity control 181, 182
  initialization and board configuration 180
  interrupt ISR 178
  libraries and constant definitions 176, 177
  lighting control 182
  project code 183
  serial feedback and calibration 183
  temperature control 180, 181

I

infrared light emitter 87
**input and output devices,
       greenhouse controller**
  about 170
  Buzzer 170
  humidity resistive sensor 170
  Motor 170
  photocell 170
  push button 170
  Relay 170
  relay, as mediator 171, 172
  Servomotor 170
  thermistor 170
inputs
  used, for sensing 77, 78
Inter-Integrated Circuit (I2C)
  about 133
  URL 133, 134
interrupts 87, 149, 150
Interrupt Service Routine (ISR)
  about 150
  CHANGE 150
  FALLING 150
  LOW 150
  RAISING 150

L

LED circuit 46, 47
LED code
  fading LED code, completing 66-68
library sound functions 113, 114
light
  programming, for sensing 101
lighting control, greenhouse controller 169
LilyPad Arduino 14-16
LilyPad board
  URL 16
LilyPad Simple board
  URL 16
LilyPad SimpleSnap board
  URL 16
LilyPad USB board
  URL 16
Linux
  drivers, installing for 32
  package, downloading for 29
  software, installing for 30
loop() function 50

M

Mac OS X
  drivers, installing for 32
  package, downloading for 28
  software, installing for 29
map() function, parameters
  fromHigh 96
  fromLow 96
  toHigh 96
  toLow 96
  value 96
Mechanical mice section
  URL 88
menus
  about 40
  Edit menu 40
  File menu 40
  Help menu 41
  Sketch menu, commands 41
message area 36
metronome
  about 124
  beatDuration variable 128
  bpm variable 128
  circuit 125
  code 126-128
  readButtons variable 128
micros() function 112
millis() function 112
modular design, greenhouse controller
  humidity control 169
  lighting control 168, 169
  manual alarm 168-170
  temperature control 168
  watering 168
momentary push buttons
  about 81
  Panel mounting 80
  Printed Circuit Board (PCB) soldering 80
motor driver circuit
  about 69
  assembled 74
  base 69
  collector 69
  completed 71
  connections, diagram 72
  diode 70
  emitter 69
  power motors 74, 75
  power source, considerations 70, 71
  resistor 70
  speed code, varying 73
  with transistor 68-70
motor speed control schematic 106
multiplatform tool 27
multiple outputs, circuit
  dealing with 53, 54
myActiveDelay() function 124

O

optical switch
  breadboard connections, diagram 90, 91
optocoupler
  about 88, 152
  infrared light emitter 87
  phototransistor 87
  using, as coin detector 89

P

package, downloading
  about 28
  for Linux 29
  for Mac OS X 28
  for Windows 28
  source code 29
panel mounting 80
phototransistor 87
PIR motion detector
  URL 87
potentiometer 106
Power over Ethernet (PoE) module
  URL 14
preflight checks 36-38
Printed Circuit Board (PCB) soldering 80
Pulse Width Modulation (PWM)
  about 10, 46
  URL 62

R

reed relay
  URL 87
Reference
  page, URL 147
  section, URL 86, 124
relay
  URL 172
revolutions per minute (rpm) 162

S

sensor
  calibrating 105
serial communication
  baud rate 133
  concepts 131, 132
  Inter-Integrated Circuit (I2C) 133
  Serial Peripheral Interface (SPI) 133
  types 133, 134
  URL 133
Serial Monitor command 41
Serial Peripheral Interface (SPI)
  about 133
  URL 133

setup() function 50
shields
  about 23
  URL 24
sketch
  preflight checks 37, 38
  uploading 38-40
Sketch menu
  Add File 41
  Import Library 41
  Show Sketch Folder 41
  Verify / Compile 41
software, installing
  about 29
  for Linux 30
  for Mac OS X 29
  for Windows 29
sound
  about 112
  connection, direct 115
  connection, through transistor 116, 117
  hardware connection 114
  library sound functions 113, 114
SparkFun
  URL 24
stopping
  versus accounting 112
switch / case control structure
  syntax 102-104
switches
  URL 87

T

tachograph project
  about 151
  interrupt, and ISR 154, 155
  interrupt tester 152, 153
  mechanical considerations 151, 152
temperature control, greenhouse controller
  about 169
  Fan 169
  Retractable roof 169
  Visual indicator 169
  Watering pump 169
time control functions 111

timer
  about 118
  coding, delays used 119, 120
  coding, without delays 121-124
  LED, blinking while waiting 121-124
  myActiveDelay() function 124
  sketch, dividing into different files 118
  tryToBlinkaLED() function 124
timing 87
tone() function
  duration parameter 113
  frequency parameter 113
  pin parameter 113
toolbar, buttons
  New 34
  Open 34
  Save 34
  Serial Monitor 34
  Upload 34
  Verify 34
transistor
  used, for connection 116, 117
tryToBlinkaLED() function 124

U

Universal Asynchronous Receiver/Transmitter (UART) 131

V

variable resistor
  connecting, to Arduino 98
voltage divider 99

W

Windows
  drivers, installing for 31
  package, downloading for 28
  software, installing for 29
  URL 32

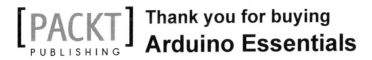

Thank you for buying
Arduino Essentials

About Packt Publishing

Packt, pronounced 'packed', published its first book, *Mastering phpMyAdmin for Effective MySQL Management*, in April 2004, and subsequently continued to specialize in publishing highly focused books on specific technologies and solutions.

Our books and publications share the experiences of your fellow IT professionals in adapting and customizing today's systems, applications, and frameworks. Our solution-based books give you the knowledge and power to customize the software and technologies you're using to get the job done. Packt books are more specific and less general than the IT books you have seen in the past. Our unique business model allows us to bring you more focused information, giving you more of what you need to know, and less of what you don't.

Packt is a modern yet unique publishing company that focuses on producing quality, cutting-edge books for communities of developers, administrators, and newbies alike. For more information, please visit our website at www.packtpub.com.

Writing for Packt

We welcome all inquiries from people who are interested in authoring. Book proposals should be sent to author@packtpub.com. If your book idea is still at an early stage and you would like to discuss it first before writing a formal book proposal, then please contact us; one of our commissioning editors will get in touch with you.

We're not just looking for published authors; if you have strong technical skills but no writing experience, our experienced editors can help you develop a writing career, or simply get some additional reward for your expertise.

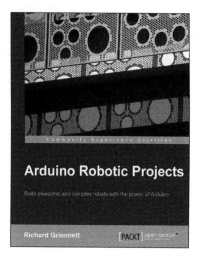

Arduino Robotic Projects

ISBN: 978-1-78398-982-9   Paperback: 240 pages

Build awesome and complex robots with the power of Arduino

1. Develop a series of exciting robots that can sail, go under water, and fly.
2. Simple, easy-to-understand instructions to program Arduino.
3. Effectively control the movements of all types of motors using Arduino.

Arduino Networking

ISBN: 978-1-78398-686-6   Paperback: 118 pages

Connect your projects to the Web using the Arduino Ethernet library

1. Learn to use the Arduino Ethernet shield and Ethernet library.
2. Control the Arduino projects from your computer using the Arduino Ethernet.
3. This is a step-by-step guide to creating Internet of Things projects using the Arduino Ethernet shield.

Please check **www.PacktPub.com** for information on our titles

Arduino Home Automation Projects

ISBN: 978-1-78398-606-4 Paperback: 132 pages

Automate your home using the powerful Arduino platform

1. Interface home automation components with Arduino.
2. Automate your projects to communicate wirelessly using XBee, Bluetooth and WiFi.
3. Build seven exciting, instruction-based home automation projects with Arduino in no time.

Internet of Things with the Arduino Yún

ISBN: 978-1-78328-800-7 Paperback: 112 pages

Projects to help you build a world of smarter things

1. Learn how to interface various sensors and actuators to the Arduino Yún and send this data in the cloud.
2. Explore the possibilities offered by the Internet of Things by using the Arduino Yún to upload measurements to Google Docs, upload pictures to Dropbox, and send live video streams to YouTube.
3. Learn how to use the Arduino Yún as the brain of a robot that can be completely controlled via Wi-Fi.

Please check www.PacktPub.com for information on our titles

Made in the USA
Columbia, SC
08 January 2019